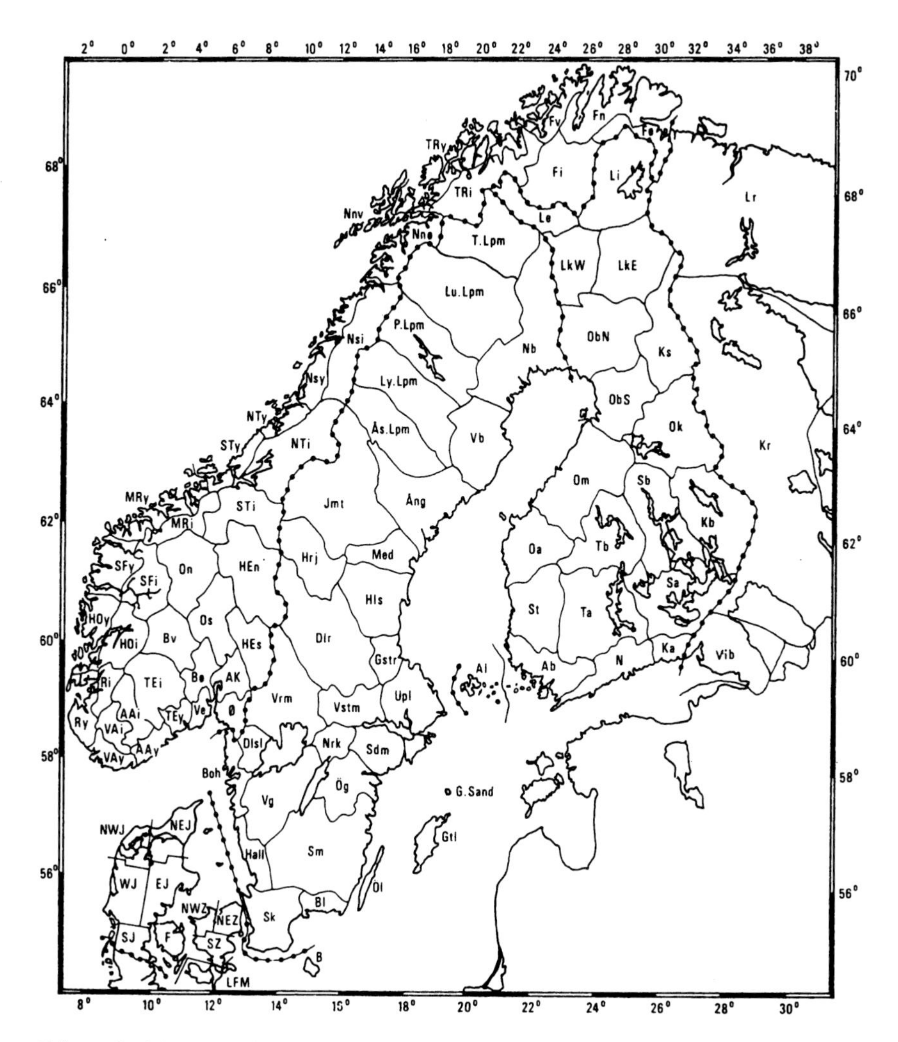

List of abbreviations for the provinces used throughout the text, on the map and in the following tables.

DENMARK

SJ	South Jutland	LFM	Lolland, Falster, Møn
EJ	East Jutland	SZ	South Zealand
WJ	West Jutland	NWZ	North West Zealand
NWJ	North West Jutland	NEZ	North East Zealand
NEJ	North East Jutland	B	Bornholm
F	Funen		

(Continued on back cover)

FAUNA ENTOMOLOGICA SCANDINAVICA
Volume 8 1979

The Formicidae (Hymenoptera) of Fennoscandia and Denmark

by

C. A. Collingwood

SCANDINAVIAN SCIENCE PRESS LTD.
Klampenborg . Denmark

Fauna entomologica scandinavica
is edited by »Societas entomologica scandinavica«

Managing editor
Leif Lyneborg

World list abbreviation
Fauna ent. scand.

Printed by
Vinderup Bogtrykkeri A/S
7830 Vinderup, Denmark

Publication date
11 June, 1979

ISBN 87-87491-28-1

Contents

Introduction

The only reference work for European Formicidae that includes descriptions of most of the species found in Denmark and Fennoscandia is that of Stitz (1939). Redefinitions of certain species, nomenclature changes, the discovery of a few additional species as well as many new distribution records have inevitably made the systematic part of that work out of date. The most recent and valuable work dealing substantially with the same fauna is that of Kutter (1977) which, although restricted formally to the species actually recorded within Switzerland, makes descriptive reference to the very few additional species that occur in Fennoscandia. Other standard works that deal with limited faunae in adjacent areas include those of Pisarski (1961, 1975) for Poland, Bolton and Collingwood (1975) for the British Isles and van Boven (1977) for Belgium. Douwes (1976) following Forsslund (1957b) has brought the Swedish list of Formicidae up to date with a key to the species worker caste. Larsson (1943) dealt with the Danish fauna but this was mainly extracted from the information given by Stitz (1939). The Norwegian fauna was listed and keyed by Holgersen (1944) and the distribution and nomenclature brought up to date by Collingwood (1974). There exists no standard work to the Finnish fauna. The present work owes much to the revisionary studies of certain genera and species groups including Yarrow (1954, 1955a) for British species allied to *Formica fusca* and *F. rufa* respectively, Wilson (1955) for members of the genus *Lasius* and extended faunal reviews such as that of Dlussky (1967) for the genus *Formica* in the USSR. Literature references are biased towards papers by Danish and Fennoscandian authors since they deal appropriately with the ecology and distribution of species occurring within the area covered by the present work.

Grateful acknowledgement is made for information, specimens and often hospitality provided by Ch. Bisgaard and Chr. Skøtt in Denmark, P. Douwes in Sweden, P. Sveum, T. Kvamme and A. Fjellberg in Norway and to R. Rosengren and H. Wuorenrinne in Finland and to many other individuals who contributed records, also Dr Astrid Løken of the University Zoology Museum, Bergen, Dr H. Waldén of the Natural History Museum, Göteborg, Dr A. Lillehammer of the Natural History Museum, Oslo, Dr W. Hackman of the Natural History Museum, Helsinki, and K.-J. Hedqvist of the Natural History Museum, Stockholm for permission to examine their local collections. All these as well as my own brief excursions to these territories have enabled me to complete as up to date a picture as possible of the distribution of the local fauna.

In addition I am grateful for permission to include certain photographs and illustrations to C. F. Jensen, Århus and to Professor D. Lee of Leeds University from a project carried through by R. I. Lawson. Above all I am deeply grateful to Peter Hurworth who drew illustrations from individual specimens of most of the Myrmicinae and some of the species of the other subfamilies. I am indebted also to Barry S. Bolton of the British Natural History Museum for suggesting certain essential textual corrections and to Leif Lyneborg for careful and helpful editing.

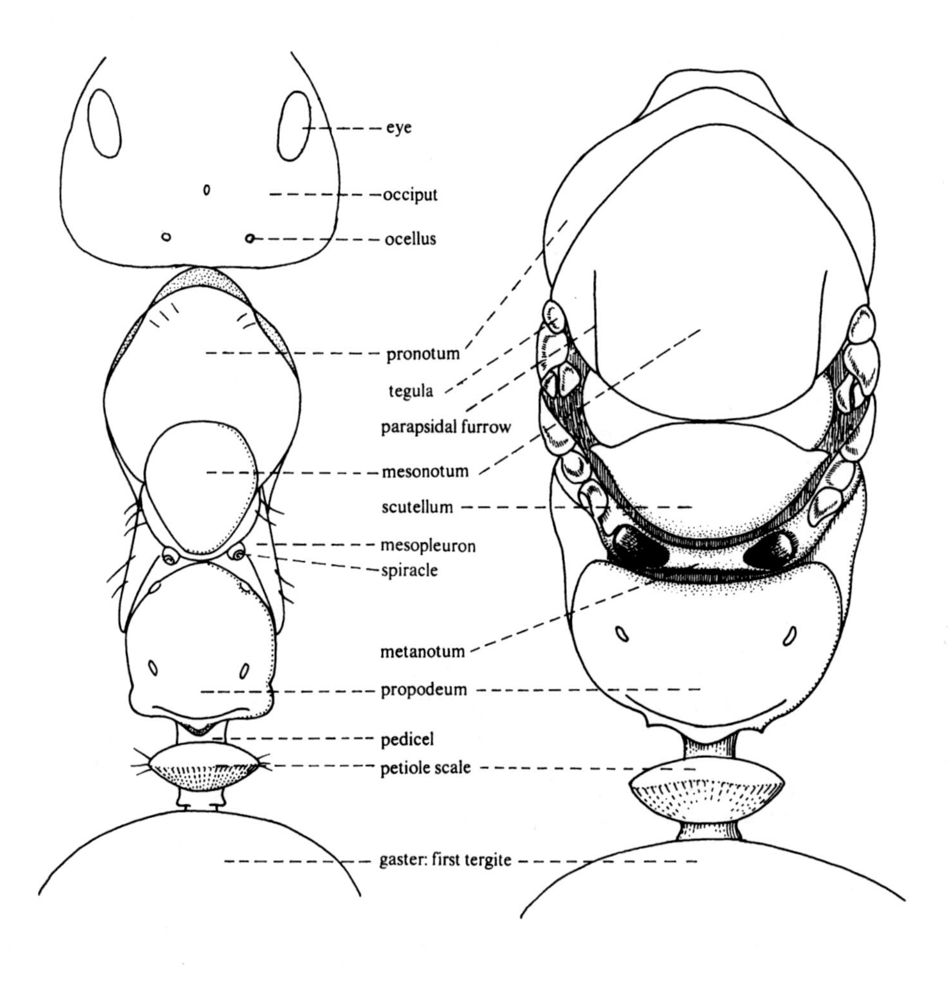

Fig. 1. Dorsal view of *Formica* sp., worker (left) and queen (right).

Diagnosis and morphology

The superfamily Formicoidea includes but one family Formicidae consisting of several subfamilies of which only Ponerinae, Dolichoderinae, Myrmicinae and Formicinae are represented in Northern Europe. All ants are characterised by the possession of a petiolus with one segment, the petiole or two segments, the petiole and postpetiole. These together with the propodeum (or epinotum of older authors) are morphologically abdominal segments shifted forward from the gaster. The worker caste which is present in the vast majority of species is entirely wingless and this feature together with the segmented petiolus and consistent social organisation distinguish the Formicidae from other aculeate Hymenoptera. The Formicidae are thought to be derived from a Tiphiid ancestor. Supporting evidence is given by Wilson, Carpenter and Brown (1967) who describe *Sphecomyrma,* a wasplike but true wingless ant and the oldest known ant fossil preserved in amber from the early Upper Cretaceous dated some 100 000 000 years ago.

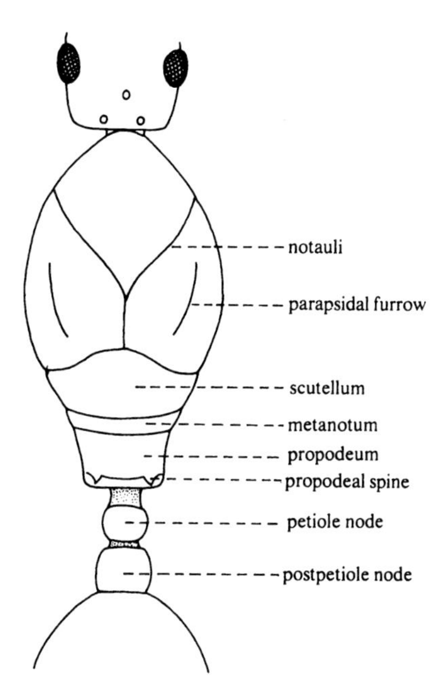

Fig. 2. Dorsal view of male of *Tetramorium* sp.

The various features referred to in the dichotomous keys and descriptions are illustrated in Figs. 1–11. The head which bears the antennae including the scape and funiculus, eyes and ocelli, is fronted by a shield or clypeus which protects the mouth parts including the labrum, labial and maxillary palps and overhangs the mandibular insertions. The mid body, or alitrunk, includes the various fused nota – pronotum, mesonotum, metanotum and propodeum. Ventrally the legs are attached by coxae to the propleuron, mesopleuron and metapleuron and include the trochanter, femur, tibia and five tarsal segments terminating in a bifurcated claw. The front tibiae, and in some

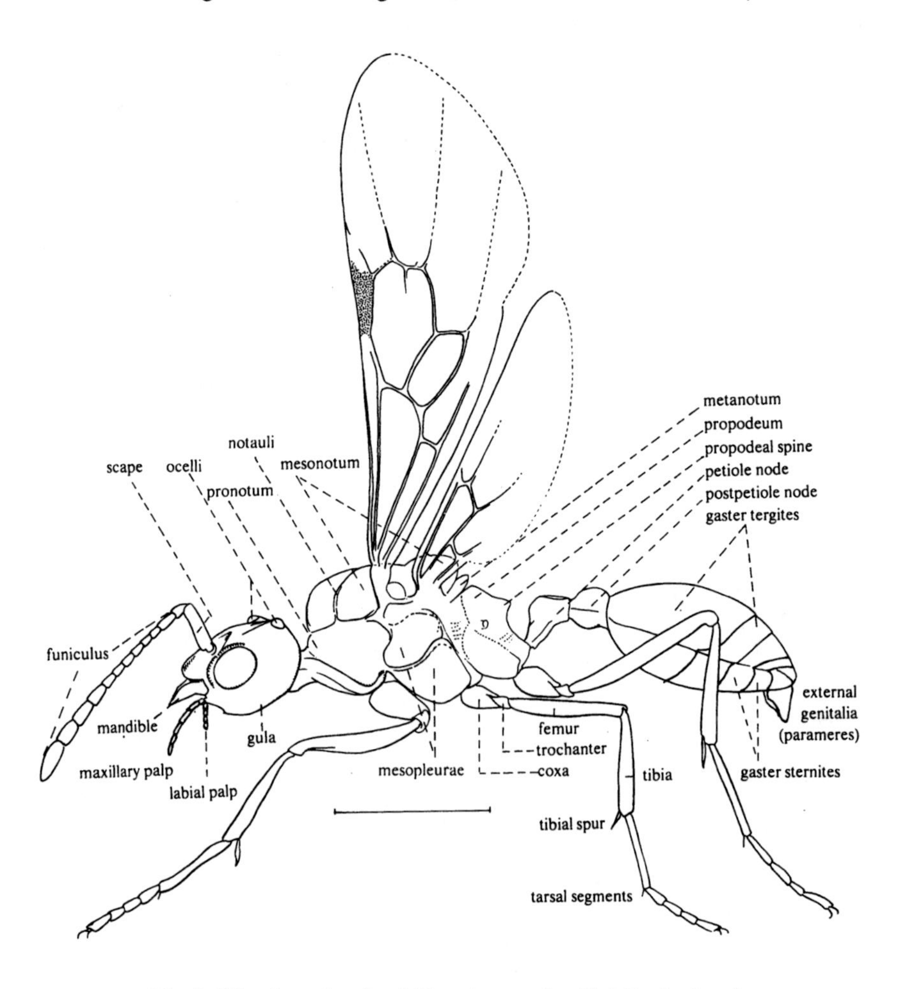

Fig. 3. Side view of male of *Myrmica rugulosa* Nyl. Scale: 1 mm.

genera the mid and hind tibiae, bear a pectinate or comblike spur. The metanotal and propodeal spiracles are usually prominent and clearly visible and the nodal or scale like petiole and the postpetiole are included in the alitrunk. The gaster consists of a number of segments including the dorsal tergites and ventral sternites ending apically in an orifice through which the sting protrudes in the female castes of Ponerinae and Myrmicinae, with a circular acidopore in Formicinae and a transverse slit in Dolichoderinae.

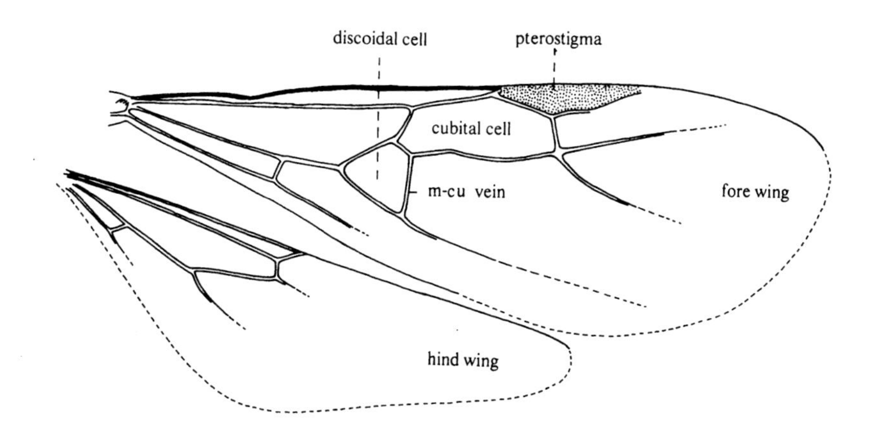

Fig. 4. Wings of *Stenamma westwoodii* Westwood.

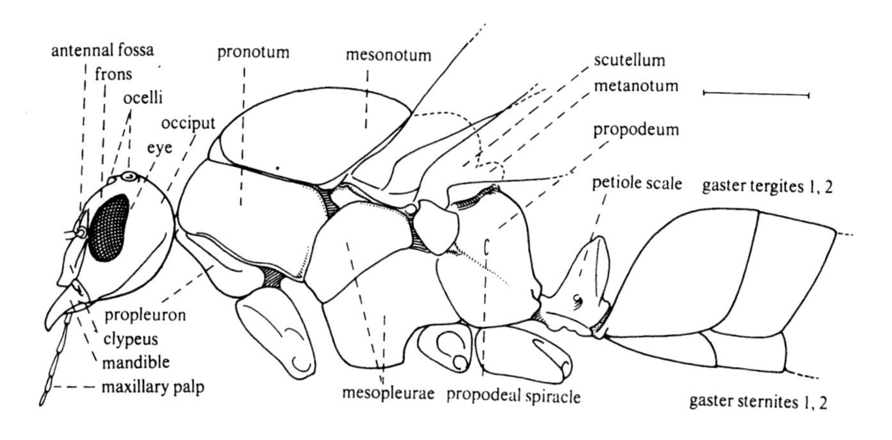

Fig. 5. Side view of male of *Formica fusca* (L.). Scale: 1 mm.

Males are distinguishable from workers and queens by an additional gaster tergite and more or less prominent genitalia. In most species the male is winged and the head relatively small compared with the female castes. Exceptions are worker-like males in the genera *Hypoponera* and *Formicoxenus* which have no wings. Males have one more funiculus segment than females except in Tetramorini, *Sifolinia* and *Hypoponera.* Queens always have wings which are quickly lost after copulation either by atrophy of the wing bases or by positive removal by biting. Worker-like queens may also occur in some genera such as *Polyergus* but these, although wingless, are easily distinguished as with winged queens from the worker by the more or less enlarged alitrunk and more prominently developed mesonotum which in most species is constituted by scutum (mesoscutum) and scutellum.

In some Myrmicinae intercaste worker/queen like forms are quite frequent with a range of wing development from wing buds to complete wings. Microgynes (= small queens) also occur, notably among some species of the genus *Myrmica.* These are also occasionally developed in *Formica* but more frequent in this genus are pseudogynes – worker-like ants with a grossly enlarged mid body. These are not functional queens and as workers their activity is of a low order and they are probably due to some distortion of the feeding rearing process. They are particularly common in *Formica sanguinea* and in members of the *F. rufa* group. Various other abnormalities occur, including male/female mosaics, queens and workers swollen or discoloured by internal mermithid parasites or having a telescoped alitrunk probably through mechanical injury during development. Many species have polymorphic workers seen in an extreme form in *Pheidole* with a sharp differentiation between major and minor workers or as in *Formica rufa,* a gradual progression from smallest to largest workers.

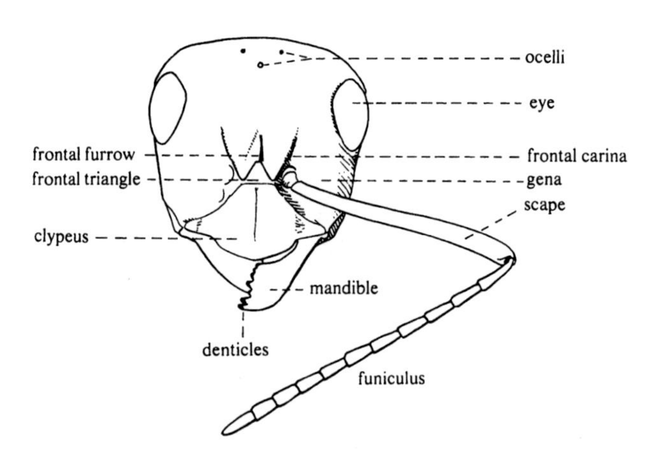

Fig. 6. Head in dorsal view of worker of *Formica fusca* (L).

The developmental stages include the eggs which are variously shaped in different species but most usually more or less oval. In addition to normal eggs, so-called trophic eggs may be produced as food for the colony; these may be infertile worker laid eggs or, in some species, of regular production by the queen or queens themselves. The eggs are coated with a sticky saliva by the nurse workers and adhere together in small packets for transport within the nest or during migration. The legless larvae are segmented, vermiform, swelling posteriorly with small but distinct heads bearing small mandibles which are functional in most species; larvae are protected by various types of hairs, thin spines and tubercules. Development time varies seasonally and with different species from a few weeks to many months. At pupation the larval skin splits dorsally and the pupal form with complete appendages and adult type segmentation is exposed. In Ponerinae and Formicinae, the pupa is usually enclosed in a protective cocoon but this is never the case with Dolichoderinae or Myrmicinae.

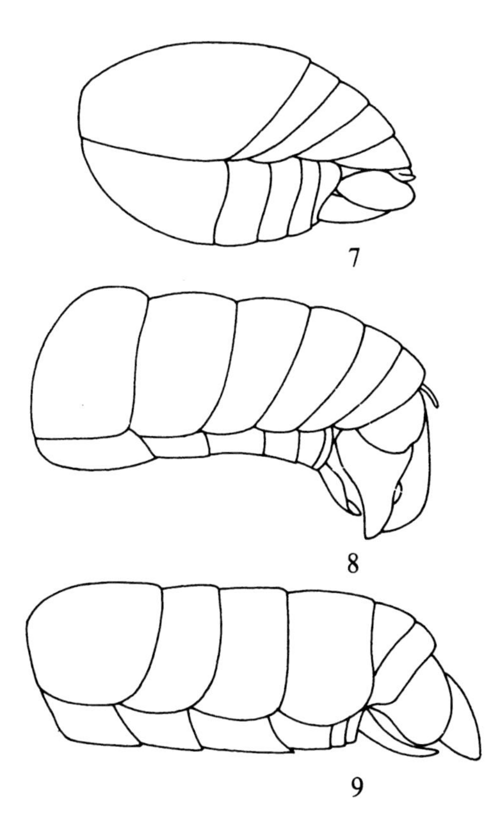

Figs. 7–9. Gaster profile of males of 7: *Myrmica ruginodis* Nyl., 8: *Formica nigricans* Emery and 9: *Tapinoma erraticum* (Latr.), showing general appearance of external genitalia.

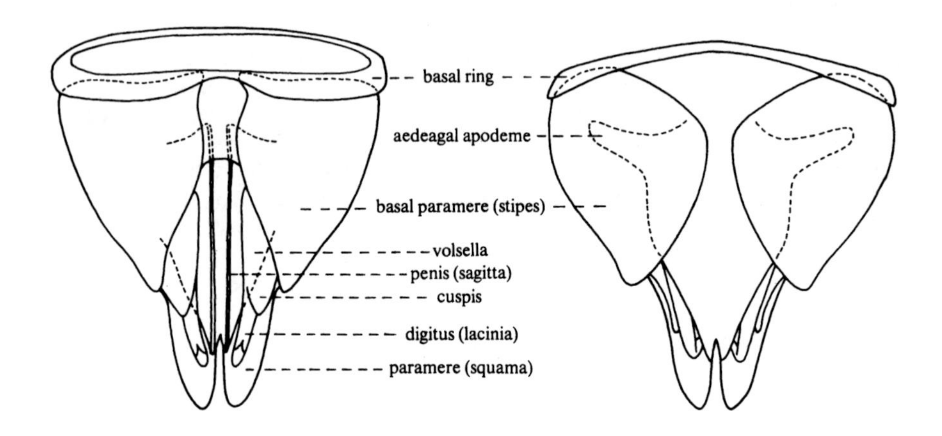

Fig. 10. Male genitalia of *Tapinoma erraticum* (Latr.), in postero-dorsal view (left) and antero-ventral view (right).

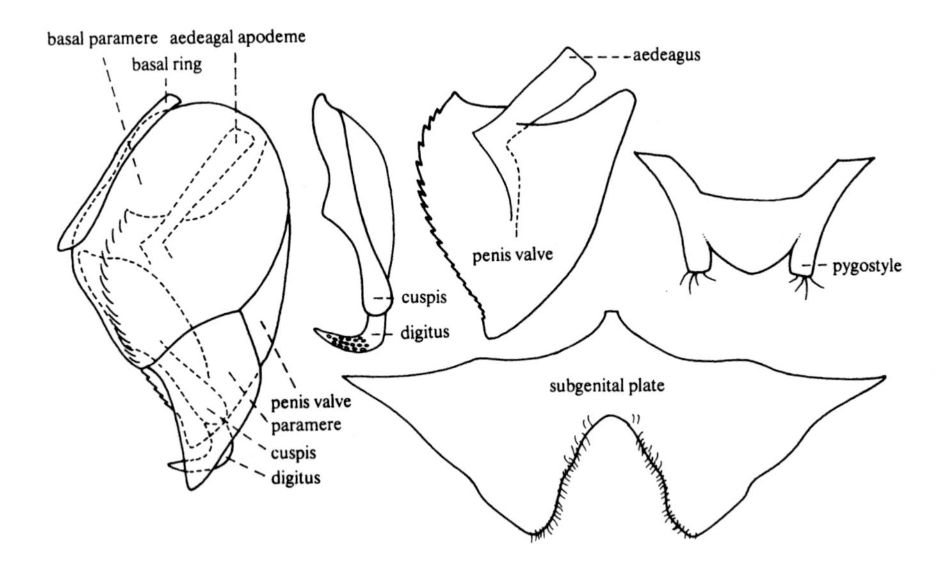

Fig. 11. Male genitalia dissected of *Tapinoma erraticum* (Latr.).

Bionomics and ecology

All ants are social. Some species form large populous colony groups of interconnected nests each with several or many queens. Others live in isolated nests consisting of few to very many workers according to species and single queens. Some are parasitic on other species and several are temporary parasites during colony initiation.

Because of their social organisation, ants can play a dominant role in particular environments. Ecological studies are various and not easily summarised, but much attention is being paid to the role of ants in the ecosystem and to food and energy budgets of common species such as *Myrmica rubra* (Petal, 1967), *Lasius alienus* (Nielsen and Jensen, 1975), and *Lasius flavus* (Pontin, 1978). In addition to predation of small arthropods, ants tend aphids and coccids for their honeydew and Pontin (1978) has shown that *Lasius flavus* may operate a more or less closed system in stable meadow habitats tending up to 30 species of aphids for their honeydew and at the same time predating surplus aphids for their protein requirement.

Populous species such as *Lasius flavus* develop large populations in suitable environments and estimated worker numbers range from 3000 to over 7000/m² (Nielsen, Skyberg and Winther, 1976). This species is often a dominant in old pasture and individual mounds develop a characteristic flora of fine herbs, grasses and moss that contrasts with the surrounding coarse grasses. Oinonen (1956) has shown that in rocky woodland exposed by burning or felling in South Finland, *Lasius flavus* by building up earth mounds in rock crevices provides shelter for pine seedlings and may thus be instrumental in natural forest regeneration.

Wood ants of the *Formica rufa* group may have a much lower worker density than many of the smaller species even in a well populated area of woodland but very large numbers of arthropods may be destroyed within the foraging area of each nest since over 30 % of the normal colony diet will consist of such prey. These ants have long been considered to play an important role in regulating foliage feeding of forest trees. Opinions differ as to the relative importance of this because of the uneven distribution of nests on the one hand and the destruction of predator species as well as tree pests on the other. Otto (1958) probably best summarises a balanced view of these activities. In some countries, notably West Germany, East Germany, Bulgaria and some states in the USSR, *Formica rufa* group species are protected, encouraged and even artificially introduced into new plantations as a means of preventing epidemic pest defoliation. Wood ant cocoons are also collected as a supplement for bird food and Wuorenrinne (1978) draws attention to a relatively massive export trade of dried ant pupae collected by local farmers in Central and Eastern Finland. Because of the abundance and relative conspicuousness of these species in the Finnish forest environment, they are also being studied as species indicators of biological environmental change consequent on degrees of suburbanisation in South Finland (Wuorenrinne and Vepsäläinen, 1976).

The density of *Myrmica* populations with their relatively small colonies is considerably lower than that of most *Lasius* species. Petal and Bremeyer (1968) estimate around 300/m² for *Myrmica rubra* in a well populated meadow but show that even at these levels, thousands of arthropods, chiefly spiders, may be destroyed per m² throughout the active period of the summer season.

Many arthropods have a more or less obligate association with ants. Thus many aphid species are never found indepentantly of ants and some show behaviour and morphological adaptation to ant tending. Insects and arthropods of several orders harbour in ant nests and a large number are ant adapted with at least part of their life cycle closely associated with their host. The majority of such ant associated arthropods are found with *Formica rufa* and its allied species and also with the more populous members of the genus *Lasius.* Behaviour ranges from true parasitism exemplified by ectoparasitic *Antennophorus* mites, through predation of the ants themselves – the beetle *Zyras humeralis* Gr. for example can destroy large numbers of *Formica rufa* workers, existence as tolerated scavengers such as beetles of the genus *Thiasophila,* and true symphily where the inquilines are fed and tended by the ants and at the same time secrete body fluid that is attractive to their hosts as food, exemplified by the beetle *Lomechusa strumosa* F. Larsson (1943) gives a useful list of European myrmecophileous inquilines including more than 300 species in several insect orders and Donisthorpe (1927) gives some detailed account of the life history of many species.

Agriculturally ants are of little importance in Northern Europe on cultivated land since all but the most common and adaptive species are destroyed by cultivation. However some species, notably *Lasius niger,* occasionally have local importance by harbouring and protecting aphid species on crop plants. Also damage to orchard blossom by biting the developing stigma may occur on a very localised scale. Ants may also act as intermediate hosts of certain cestode and trematode parasites of livestock. These have some importance in Central Europe and their bibliography and occurrence have been summarised by Passera (1975).

Domestically ants may be a nuisance as exemplified by the seasonal invasion of larders and kitchens by *Lasius niger* in particular, by the much more localised invasion and destruction of timber in outhouses and even domestic premises by *Lasius fuliginosus* or *Lasius brunneus* in Southern Fennoscandia. More serious is the invasion and establishment of cosmopolitan or introduced species into heated domestic buildings such as high rise flats, bakeries and hospitals. The minute *Monomorium pharaonis* is one of the worst examples of this, and Eichler (1976) summarises the occurrence of this species in North and Central Europe and the various harmful effects that can occur in hospitals by transmission of infective disease and by irritation. Several species of temperate or subtropical origin may find their way by importation in plant material into heated glasshouses. In Northern Europe such warm adapted species cannot survive for long outdoors but one species in particular, *Iridomyrmex humilis,* has successfully invaded and established itself in large polycalic colonies along the Mediterranean coast to the detriment of local indigenous species.

Distribution and faunistics

The fauna of these northern territories include a number of species of differing ranges. Because of the mixture of arctic, boreal and Central European species no clearcut distinction can be made between a northern European and a Central or Southern European fauna. Thus the arctic Siberian species *Formica gagatoides* ranges southward into the mountains of South Norway within the same latitude as such southern species as *F. cinerea* and *Lasius meridionalis* while in Sweden, rare or uncommon Central European species including *Lasius bicornis, L. carniolicus* and *Leptothorax corticalis* have been recorded very far to the north of their main ranges. Similarly *Camponotus fallax* and *C. vagus* which are common in South Europe extend sparsely and very locally into Poland and North Germany and still exist in small isolated pockets in Central Sweden and Southwest Finland respectively. The only known endemic species is *Formica suecica*

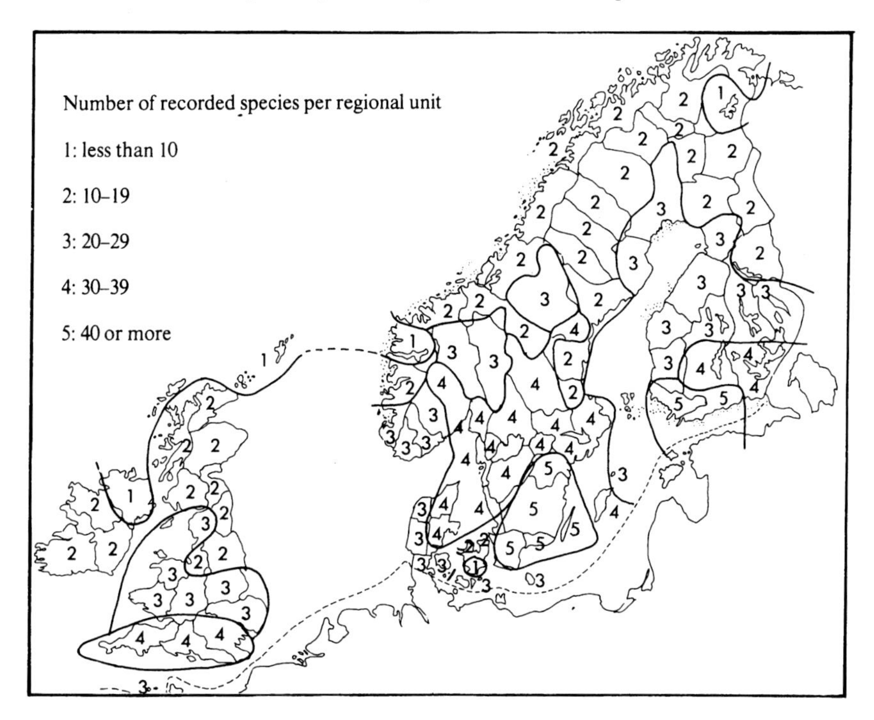

Fig. 12. Species abundance in the British Isles, Denmark and Fennoscandia.

which is not found outside Norway, Finland and Sweden where it is distributed locally and rather sporadically as far north as latitude 70°, except in Estonia in the Baltic States. To sum up, the fauna can be characterised as mainly wide ranging northern boreal Eurosiberian with a number of south boreal species and a mixture of much more local or rare species with main ranges in South Central Europe.

Fig. 12 shows species abundance within the regions of the various territories including the British Isles. South Zealand is anomalously low in species but the general trend follows many distribution patterns with isolines running southwest to northeast. The most productive areas are to be found in South Sweden and South Finland. The British Isles which are included in this survey are of interest in that species such as *Myrmecina graminicola, Stenamma westwoodii* and *Diplorhoptrum fugax* are widely distributed in South England but very local indeed in Fennoscandia which has by contrast a much greater abundance of *Formica* species. Table 1 gives the number of species in each recorded genus from the British Isles, Denmark and Fennoscandia and illustrates

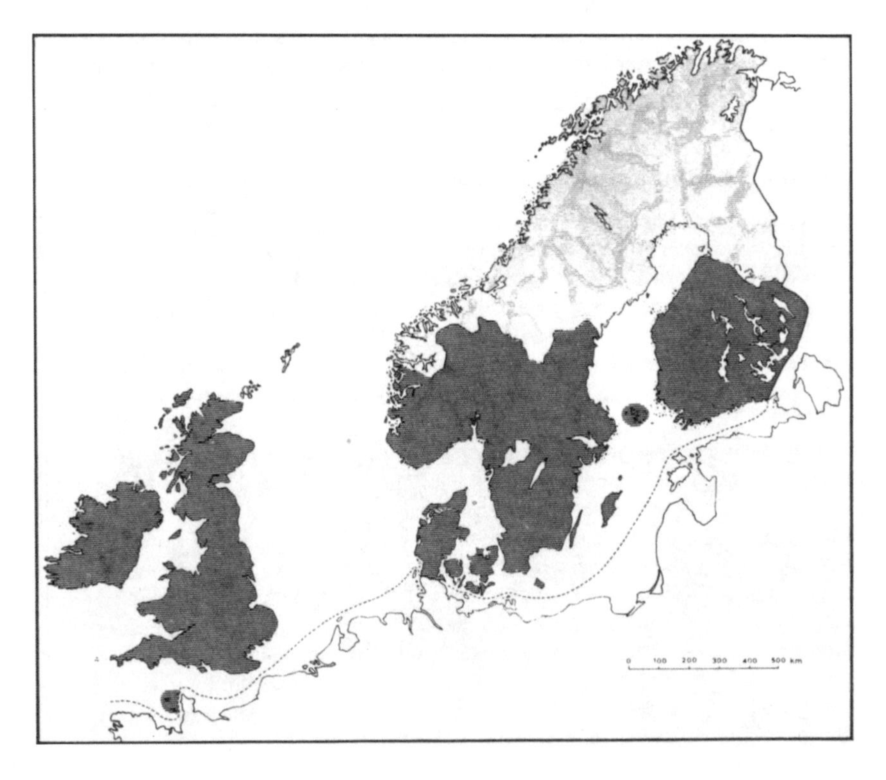

Fig. 13. Area units partitioned at 50 % similarity level (from Baroni Urbani & Collingwood, 1977).

this disparity as well as showing the complete absence of *Camponotus* species at least one of which is abundant throughout Fennoscandia.

An analysis of the ant fauna in relation to geographical and environmental parameters has been attempted by Baroni Urbani and Collingwood (1977). Fig. 13 shows the regions partitioned into areas of 50 % species similarity. Here it is seen that despite the obvious differences between the British Isles and Fennoscandia the whole area divides clearly into two zones with the whole of the British Isles, Denmark and Southern Fennoscandia differentiated from Northern Fennoscandia. Conclusions from this paper (op. cit.) suggested that the main determinants influencing species abundance included bright sunshine, July temperature, relative oceanicity and length of growing season. Twelve sets of species groups clustering at a similarity level of around 50 %

Table 1 : Number of species

Genus	British Isles*	Denmark	Sweden	Norway	Finland
Hypoponera	1	1	1	1	1
Ponera	1	–	–	–	–
Tapinoma	1	–	1	–	–
Myrmica	9	10	9	7	10
Sifolinia	1	–	1	1	1
Stenamma	1	1	1	1	–
Diplorhoptrum	1	–	1	–	–
Myrmecina	1	1	1	–	–
Leptothorax	5	4	7	4	3
Formicoxenus	1	1	1	1	1
Harpagoxenus	–	1	1	1	1
Anergates	1	1	1	–	–
Strongylognathus	1	–	1	–	–
Tetramorium	1	1	1	1	1
Camponotus	–	2	4	2	3
Lasius	9	8	9	8	8
Plagiolepis	1	–	–	–	–
Formica	11	18	20	19	18
Polyergus	–	–	1	–	–
Total	46	49	61	46	47

*Including Channel Isles

Table 2: Species list

Indigenous	Introduced
Ponerinae	
Hypoponera punctatissima (Roger)	
(*Ponera coarctata* (Latreille))	
Dolichoderinae	
Tapinoma erraticum (Latreille)	*Iridomyrmex humilis* (Mayr)
	Tapinoma melanocephalum (Fabricius)
Myrmicinae	
Myrmica gallieni Bondroit	
Myrmica lobicornis Nylander	
Myrmica rubra (Linné)	
Myrmica ruginodis Nylander	
Myrmica rugulosa Nylander	
Myrmica sabuleti Meinert	
Myrmica scabrinodis Nylander	
Myrmica schencki Emery	
Myrmica specioides Bondroit	
Myrmica sulcinodis Nylander	
(*Myrmica hirsuta* Elmes)	
Sifolinia karavajevi (Arnoldi)	
Stenamma westwoodii Westwood	*Pheidole megacephalum* (Fabricius)
Diplorhoptrum fugax (Latreille)	*Monomorium pharaonis* (Linné)
Myrmecina graminicola (Latreille)	*Crematogaster scutellaris* (Olivier)
Leptothorax acervorum (Fabricius)	
Leptothorax muscorum (Nylander)	
Leptothorax nylanderi (Förster)	
Leptothorax corticalis (Schenck)	
Leptothorax interruptus (Schenck)	
Leptothorax tuberum (Fabricius)	
Leptothorax unifasciatus (Latreille)	
Formicoxenus nitidulus (Nylander)	
Harpagoxenus sublaevis (Nylander)	
Anergates atratulus (Schenck)	
Strongylognathus testaceus (Schenck)	*Tetramorium bicarinatum* (Fabricius)
Tetramorium caespitum (Linné)	*Tetramorium simillimum* (Smith)

Formicinae

Camponotus fallax (Nylander)
Camponotus vagus (Scopoli)
Camponotus herculeanus (Linné)
Camponotus ligniperda (Latreille)
Lasius flavus (Fabricius)
Lasius alienus (Förster)
Lasius brunneus (Latreille)
Lasius niger (Linné)
(*Lasius emarginatus* (Olivier))
Lasius fuliginosus (Latreille)
Lasius umbratus (Nylander)
Lasius meridionalis (Bondroit)
Lasius bicornis (Förster)
Lasius mixtus (Nylander)
Lasius carniolicus (Mayr)
(*Plagiolepis vindobonensis* Lomnicki)
Formica fusca Linné
Formica gagatoides Ruzsky
Formica lemani Bondroit
Formica transkaucasica Nasonov
Formica cinerea Mayr
Formica cunicularia Latreille
Formica rufibarbis Fabricius
Formica exsecta Nylander
Formica foreli Emery
Formica forsslundi Lohmander
Formica pressilabris Nylander
Formica suecica Adlerz
Formica uralensis Ruzsky
Formica sanguinea Latreille
Formica truncorum Fabricius
Formica rufa Linné
Formica polyctena Förster
Formica aquilonia Yarrow
Formica lugubris Zetterstedt
Formica pratensis Retzius
Formica nigricans Emery
Polyergus rufescens (Latreille)
Paratrechina longicornis (Latreille)
Paratrechina vividula (Nylander)

were found to represent the major distribution pattern of the area. No evidence could be found for postulating pre-glacial relicts and most of the present day species distribution could be accounted for by general movement from the southeast and south. Table 2 lists all the species discussed in the present work including a few not recorded from Denmark or Fennoscandia but widely distributed in either or all of the Channel Islands, Low Countries, North Germany and Poland. In addition to native species found occurring naturally outdoors some twelve species of cosmopolitan or southern origin are described and keyed since they are frequently introduced and in some cases have long been established in heated premises in one or other of the territories included within the scope of this work. While distributions and species lists are as complete as possible on present information, the recent addition of such species as *Sifolinia karavajevi, Myrmica gallieni* and *M. specioides* to the Fennoscandian fauna and the extension of known ranges of the less common species suggest that further collecting will continue to be profitable and should be encouraged.

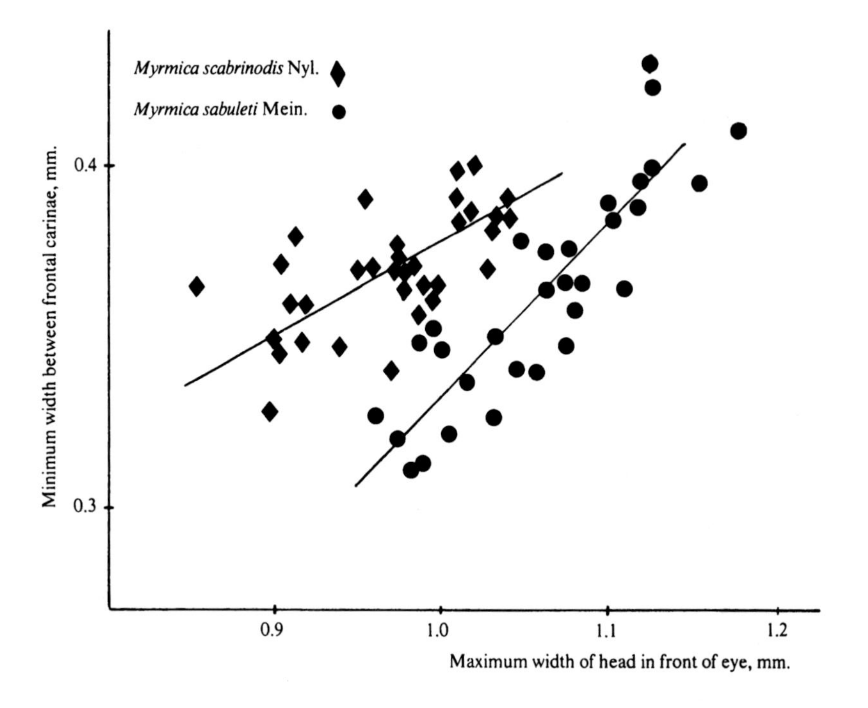

Fig. 14. Head measurements in workers of *Myrmica scabrinodis/sabuleti.*

Nomenclature and systematics

Most European authors have used formal subgeneric taxa especially in *Lasius* and *Formica* as in Kutter (1977). Apparent clear distinctions in useful characters between species groups in local faunas tend to merge or to be imprecise when the particular genus is studied on a world basis, and following the general modern trend as in Bolton and Collingwood (1975) subgeneric names are not used here. Older authors including Stitz (1939) also used terminology implying hybrid origin between related species. Occasional hybrids may well occur in nature and have been authenticated in at least two genera *Solenopsis* (Hung and Vinson, 1977) and *Acanthomyops* (Wing, 1968) in

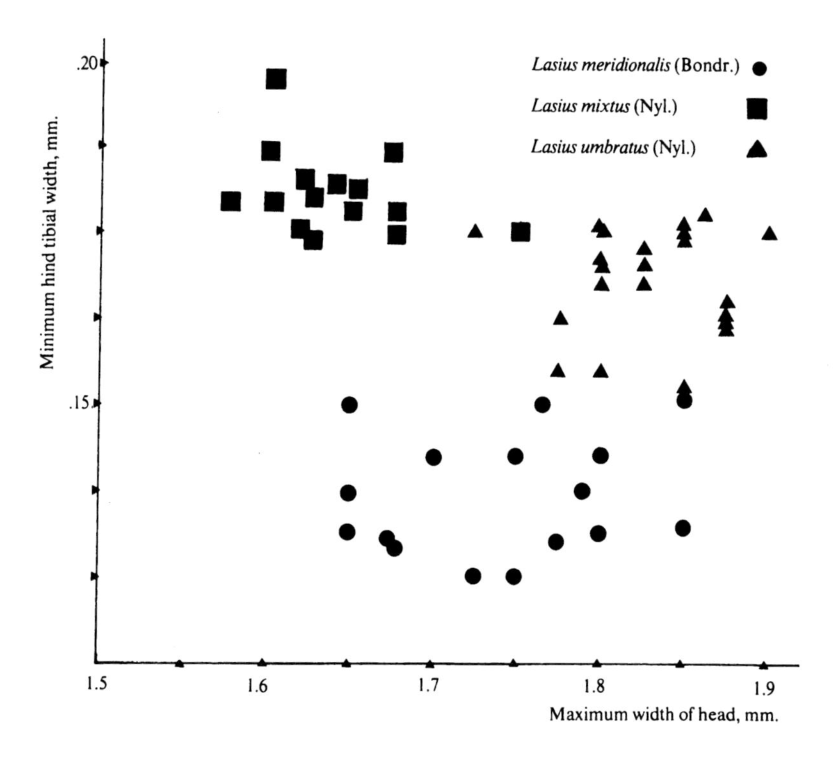

Fig. 15. Correlation between width of hind tibia and head width in the queens of three species of *Lasius*.

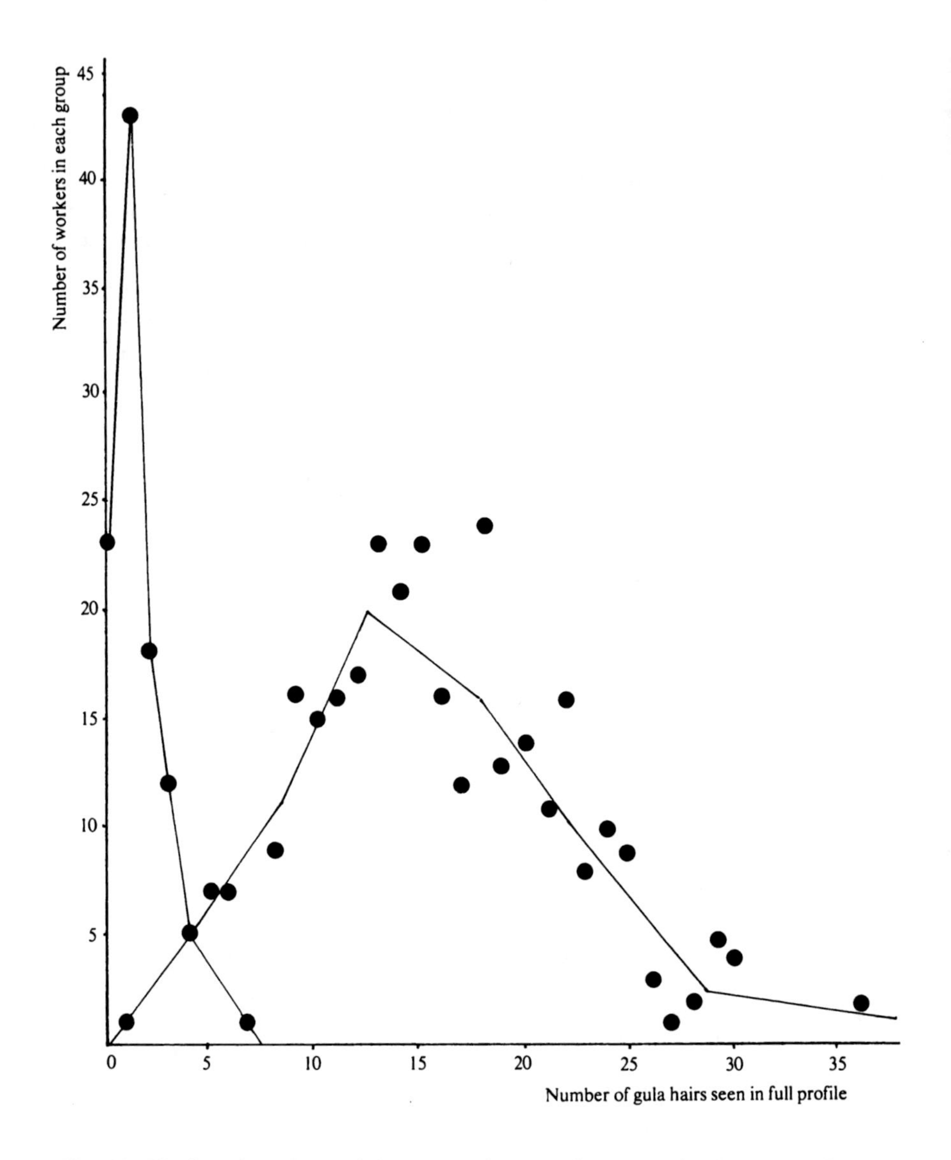

Fig. 16. Pilosity of workers of *Formica polyctena/rufa.* Range for *Formica polyctena* Förster: 0–7 hairs per worker, mean 1.12. Range for *Formica rufa* L.: 2–40, mean 18.2.

America but they are evidently rare and probably not easily recognised. Names for infraspecific forms are not used in the present work although in some groups, notably *Formica rufa* and allied species and also *Lasius umbratus*, variability in useful discriminating characters can make certain identification very difficult. In the *F. rufa* group such variability in morphology as well as in life style suggests that this complex is still undergoing evolutionary differentiation.

In such genera as *Myrmica* and *Leptothorax*, shape of structures often offer the most certain means of discrimination between similar species. In *Formica* and *Lasius* where the basis shape tends to be similar within each species group, the presence, absence or relative abundance of standing hairs on particular parts of the body may provide the most useful and easily perceived differences. Often species differentiation can be aided by the use of measurements and this is illustrated in figs. 14–16. Thus *Myrmica sabuleti* and *M. scabrinodis* may be difficult to separate on the scape character in some instances but by plotting regression lines as in fig. 14 it can be seen that only in the upper size range of workers is there likely to be overlap between the species. Similarly in fig. 15 three members of the *Lasius umbratus* complex could be separated in most cases by the relative width of the hind tibia. *Formica rufa* and *F. polyctena* are separable only on the basis of relative abundance of body hairs but fig. 16 shows that although there is overlap, these supposed species can be separated in the majority of instances by the number of gula hairs in the worker caste. Such measurements may be employed in a more complex manner and subjected to computer analysis to demonstrate degrees of separation and relatedness in similar species. Kutter (1976, 1977) has used a code diagnosis for easily perceived characters to show such relationships but such codes, although useful in defining a species, may tend to obscure the really significant characters or character which can be used to separate like species. Statistical techniques pioneered in a simple form by Brian and Brian (1949) for discrimination between *Myrmica rubra* and *M. ruginodis* and by Wilson (1955) for members of species groups in *Lasius* are in process of rapid development with more complex analyses now becoming practicable with modern computers. Thus Elmes (1978) in a morphometric study using a number of simple linear measurements of three similar *Myrmica* species found canonical multivariate analysis to give good separation with minimal overlap.

Finally there are recent developments in biochemical analyses of the complex of components of exocrine glands that, encouragingly, have been found to support morphological differences between similar species in such genera as *Myrmica*. The analysis of mandibular gland secretions or pheromones has been reviewed by Blum (1969, 1970), one of the pioneers of this type of work and investigations on these and also the glandular components of venom apparatus in relation to both behaviour response and species relationships are in progress at many centres. Electrophoretic studies of enzyme patterns so far conducted on members of the genus *Formica* by Pamilo, Vepsäläinen and Rosengren (1975) suggest that this type of biochemical analysis may have more value in determining relationships between species groups than in demonstrating clear differences between similar species.

It is emphasised however that both morphometric studies and biochemical analyses cannot be expected to replace but rather to support the more easily seen morphological qualitative differences used in conventional ant taxonomy. In the present work complete descriptions of each species are not given but attention is drawn to those features that are considered to be most useful for quick identification. With good lighting most species may be easily separated at low magnifications from x15 to x40 but electroscan microscopy using much higher magnifications can show important differences in microsculpture as are illustrated by figs. 169–175.

Collecting, preserving and keeping

For collecting purposes ants may be killed in a relaxed condition by ethyl acetate vapour for setting within a few days. For long storage, ants may be kept in 70 % alcohol but preferably to avoid brittleness and some discolouration, in a mixture containing five parts acetic acid, 35 parts absolute alcohol and 60 parts of an aqueous solution of 0.02 % mercuric perchloride. Ants are normally set by attachment across the point of a triangular card for ready examination of all surfaces.

There are many devices for keeping ant colonies in captivity; the essentials are the provision of a range of humidity which can be arranged by central placement of a wad of cotton wool fed by a water soaked wick, for northern species a temperature range between 17°C and 22°C, darkness of the brood chamber when not under observation and a suitable diet which may include Drosophilid pupae for protein and some form of sugar solution or honey. The non-stick substance polytetrafluoroethylene painted around the vertical rims of glass or plastic containers effectively prevents escape. Brian (1977) in a general account of British species describes some useful techniques.

Key to subfamilies of Formicidae

Females

1 Pedicel with two nodal segments – the petiole and postpetiole (Fig. 73) **Myrmicinae** (p. 36)

– Pedicel with one segment – the petiole, either a node or a scale 2

2 (1) Gaster distinctly constricted between first and second segments; sting present and prominent (Fig. 18) .... **Ponerinae** (p. 29)

– Gaster without constriction between first and second segments; sting absent 3

3 (2) Anal aperture in the form of a transverse slit; four gastral tergites visible in dorsal view. North European species lack standing hairs on dorsum of alitrunk and head posterior to clypeus (Fig. 22) .... **Dolichoderinae** (p. 32)

– Ventral apex of hypopygium developed into a conical structure with a circular opening; five gastral tergites visible in dorsal view. Some standing hairs always present on dorsum of head and on dorsum or sides of alitrunk (Fig. 124) **Formicinae** (p. 85)

Males

1 Pedicel with two nodal segments – the petiole and postpetiole (Fig. 3)
Myrmicinae (p. 36)
– Pedicel with one segment – the petiole, either a node or a scale 2
2 (1) Gaster distinctly constricted between first and second segment (Fig. 19)
Ponerinae (p. 29)
– Gaster without constriction between first and second segment 3
3 (2) Five gastral tergites visible in dorsal view. North European species lack standing hairs on alitrunk or head posterior to clypeus (Fig. 21) .... **Dolichoderinae** (p. 32)
– Six gastral tergites visible in dorsal view. Standing hairs always present on dorsum of head and alitrunk (Fig. 133) .... **Formicinae** (p. 85)

SUBFAMILY PONERINAE LEPELETIER

In all castes the petiole is in the form of a thickened node which may be simple or variously ornamented. All tibiae have pectinate spurs. Wings where present have two closed cubital cells and one discoidal cell. Stings in female castes are well developed. Pupae are enclosed in cocoons.

This subfamily includes about 70 genera, most of which occur in the tropics and subtropics. In North Europe only two genera are represented in the tribe Ponerini, each with a single species.

Keys to genera and species of Ponerinae

Workers

1 Maxillary palp one segmented. Petiole without fenestrate or posterior dentate ventral process. Frontal furrow is continued as a fine median line to near occipital border. Antennal scape does not reach to the posterior border of head (Fig. 17)
1. Hypoponera punctatissima (Roger)
– Maxillary palp two segmented. Petiole with ventral paired posterolateral teeth seen in side view as a minute, backward directed angular process. Scape reaches posterior border of head (Fig. 18) .... *Ponera coarctata* (Latreille)

Queens

1 Maxillary palp one segmented. Petiole without dentate ventral process. Eyes situated at sides of head near to clypeal border... *1. Hypoponera punctatissima* (Roger)
– Maxillary palp two segmented. Petiole with ventral paired posterolateral teeth. Eyes at sides of head are distant from the clypeal border by the length of the longest eye diameter .... *Ponera coarctata* (Latreille)

Males

1 Apterous, worker-like. Antennae twelve segmented with scape as long as eight following funiculus segments. Maxillary palp one segmented. Pygidial spine absent
1. Hypoponera punctatissima (Roger)
– Wings present. Antennae with 13 segments; scape short not longer than first following funiculus segment which is swollen. Maxillary palp two segmented. Pygidial spine present (Fig. 19) .... *Ponera coarctata* (Latreille)

Genus *Hypoponera* Santschi, 1938

Hypoponera Santschi, 1938: 79; Taylor, 1967: 9, stat. n.
Type-species: *Ponera abeillei* André, 1881.

Hypoponera was split off by Santschi (1938) as a subgenus of *Ponera* and was subsequently given full generic status by Taylor (1967). It is characterised by the one segmented maxillary palp and absence of ventral ornamentation on the petiole. The genus includes a large number of small hypogoeic species occurring throughout the tropics with four species in South Europe of which only one occurs north of latitude 47°.

1. ***Hypoponera punctatissima*** (Roger, 1859)
Fig. 17.

Ponera punctatissima Roger, 1859:246.

Worker. Reddish yellow to dark brown; alitrunk and gaster thickly pubescent, finely and closely punctured. Antennae with 12 segments gradually broadening to an indefinite club; scapes do not reach posterior border of head. Frontal furrow continued as a fine line to near occipital margin. Eyes minute, set forward close to mandibular insertions. Mandibles with 3–4 strong teeth towards apex and numerous smaller denticles posteriorly. Ventral lobe of petiole simple without tooth-like process: Length: 2.5–3.2 mm.

Queen. As worker but larger with more massive alitrunk; wings present in immature, unfertilised individuals. Eyes and ocelli visible at x10 magnification. Length: 3.5–3.8 mm.

Male: Apterous, worker-like but with thinner pubescense and brighter appearance. Antenna terminates in a distinct club with scape about as long as eight following segments. No pygidial spine. Length: 3.4–3.6 mm.

Distribution. Denmark, Fennoscandia and the British Isles, recorded locally and sporadically. – Range: cosmopolitan; widely distributed throughout Europe, the tropics and subtropics.

Biology. This species is often imported with plant material. However, it has long been resident in North Europe and head capsules presumed to be of this species have been recorded from sewage mud deposited about 1500 years ago in North England. Most recorded occurrences are from heated premises such as bakehouses and conservatories. However, colonies have been recorded outside in England, Ireland, Denmark, Norway and Finland from fermenting rubbish dumps, waste tips, sawdust heaps and deep mines away from buildings. Queens and sometimes workers have also been captured individually by general herbage sweeping or in woodlands. Occurrences in Denmark and Fennoscandia have been summarised by Skøtt (1971). Colonies are often populous and many alate queens may be produced to fly out during August and September. The apterous males remain in the nest. This species, as with most Ponerini, is mainly carnivorous on small arthropods.

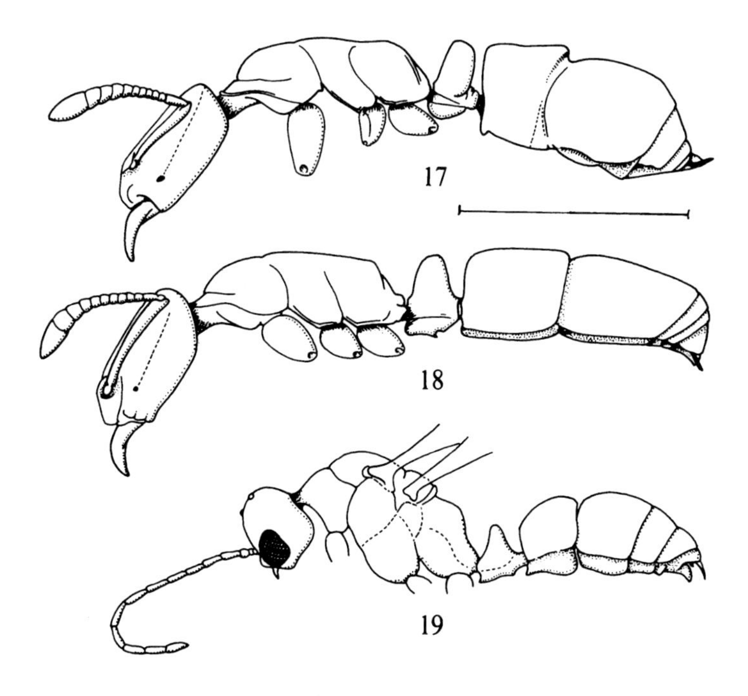

Fig. 17. *Hypoponera punctatissima* (Roger), worker in profile.

Figs. 18, 19. *Ponera coarctata* (Latr.). – 18: worker in profile; 19: male in profile. Scale: 1 mm.

Genus *Ponera* Latreille, 1804

Ponera Latreille, 1804:178.
Type-species: *Formica coarctata* Latreille, 1802b.

This genus was radically revised by Taylor (1967) who showed that *Ponera* is distinguished from the otherwise similar *Hypoponera* by the consistent presence of subpetiolar tooth-like ornamentation as well as the two segmented maxillary palps. It includes about 30 small hypogoeic species mainly distributed in the Indo-Australian region but with representatives throughout the north temperate zone of which one only occurs in North Europe.

Ponera coarctata (Latreille, 1802)
Figs. 18, 19.

Formica coarctata Latreille, 1802:65.

Worker. Light to dark brown with sparse pubescence but numerous body hairs especially on the gaster. Head more closely punctured than alitrunk; mandibles with four strong teeth towards apex and smaller indistinct denticulae posteriorly. Eyes are minute and often indistinct. Ocelli absent. Length: 3.0–3.5 mm.

Queen. Similar to worker but larger with easily visible eyes and ocelli. Length: 4.0–4.5 mm.

Male. Shining black, alate; pygidium terminating in a downcurved spine; scape shorter than second funiculus segment. Alitrunk high with arched scutellum and projecting post-scutellum. Length: 3.4–3.8 mm.

Distribution. Not recorded from Denmark or Fennoscandia. Present locally in South England and Wales and in the Channel Isles. – Range: throughout Central and South Europe from Portugal to the Caucasus and from North Africa to the Netherlands.

Biology. This is an inconspicuous species of slow movement, mainly carnivorous, living in small nests with two or three queens and 12 to 35 workers. Nests are found under stones or moss in broken stony ground, banks or crumbling cliffs and among flints in open woodland. Alates occur during August and September and have been caught by sweeping hedgerows in late summer.

SUBFAMILY DOLICHODERINAE FOREL

This subfamily includes fifteen genera with the greatest number of species distributed through the tropics. The sting is vestigial or absent but many species have poison glands that excrete a repellant fluid through the anal orifice. The gaster has a reduced number of segments compared with Formicinae and Ponerinae of which four only are visible in dorsal view in the female castes and five in the males. Pupae are not protected by cocoons. The two genera treated here both belong in the tribe Tapinomini.

Keys to genera of Dolichoderinae

Queens and workers

1 Petiole scale well developed, not obscured by overhanging gaster; front border of clypeus convex and entire (Fig. 20) .... *Iridomyrmex* Mayr (p. 33)
- Petiole a small node overhung by first gastral segment; front border of clypeus straight, incised or concave (Fig. 22) .... *Tapinoma* Förster (p. 34)

Males

1 Petiole scale well developed. Scape short, not reaching occipital margin *Iridomyrmex* Mayr (p. 33)
- Petiole a reduced node. Scape long, overreaching occipital margin *Tapinoma* Förster (p. 34)

Genus *Iridomyrmex* Mayr, 1862

Iridomyrmex Mayr, 1862:702.
Type-species: *Formica detecta* Smith, 1858.

Workers monomorphic; antennae 12 segmented, maxillary palps 6-, labial palps 4-segmented. Mesopropodeal impression deep and distinct. Petiole a low forward inclined scale. Queens considerably larger than workers; fore-wings with a closed radial cell, 2 closed cubital cells and one discoidal. Male not larger than worker, antennae filiform, 13 segmented with scape shorter than second funiculus segment; mesonotum relatively massive overhanging pronotum and part of head; wings with one closed cubital cell. Gaster considerably smaller than alitrunk; genital armature not conspicuously enlarged.

This genus has the largest number of species in Australasia with a few in South America of which one cosmopolitan species has become well established in South Europe.

Iridomyrmex humilis (Mayr, 1868)
Figs. 20, 21.

Hypoclinea humilis Mayr, 1868:164.

Worker. Pale bronze to bronze black with a deep mesosomal constriction giving a waisted effect accentuated by the high propodeum. Length: 2.2–2.8 mm.

Queen. Colour as worker; alitrunk enlarged and high relative to width. Length: 4.5 mm.

Male. Bronze black; head very small; alitrunk high and arched. Length: 2.5–3 mm.

Biology. This species was introduced into Europe from South America and has become an established and notorious pest in the Mediterranean area, developing very populous multi-queened colonies along the coast. It is sometimes brought into North

Europe with plant materials and occasionally colonises heated premises. It does not appear to be able to establish outside in northern latitudes but is present and said to be increasing in the Channel Islands.

Genus *Tapinoma* Förster, 1850

Tapinoma Förster, 1850:43.
Type-species: *Formica erratica* Latreille, 1798.

Workers monomorphic but size in some species variable. Antennae 12 segmented; palp formula 6, 4. Clypeus with front border truncate or incised. Petiole a small inconspicuous node overhung by the first gastral tergite. Queens as workers but with enlarged flattened alitrunk. Males have filiform thirteen segmented antennae with long scapes overreaching the occipital border. Head large relative to alitrunk. Genital armature with large prominent stipes.

This is a world-wide genus with most species found in the Indo-Australian and Ethiopian regions. There are several palaearctic species of which only one is indigenous north of latitude 50°. One or more of the smaller tropical species are cosmopolitan and sometimes establish in heated premises.

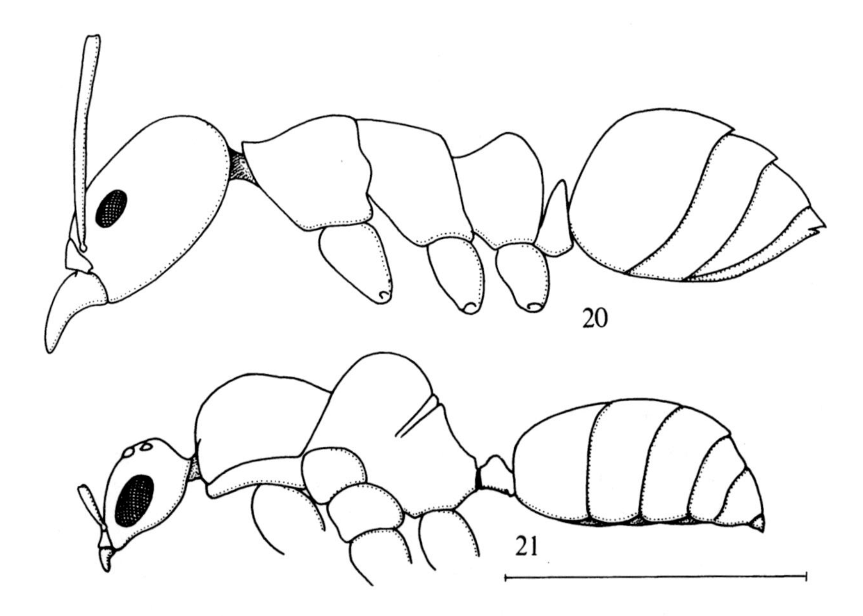

Figs. 20, 21. *Iridomyrmex humilis* (Mayr). – 20: worker in profile; 21: male in profile. Scale: 1 mm.

Keys to species of *Tapinoma*

Queens and workers

1 Median portion of anterior margin of clypeus incised; colour evenly dark brown to black .... 2. *erraticum* (Latreille)
- Median portion of anterior margin of clypeus entire; bicoloured with dark head and promesonotum contrasting with pale gaster, propodeum and appendages *melanocephalum* (Fabricius)

Males

1 Appendages and part of gaster very pale; length less than 2.5 mm *melanocephalum* (Fabricius)
- Appendages and gaster unicolorous dark with rest of body; length 3.5 mm or greater .... 2. *erraticum* (Latreille)

2. ***Tapinoma erraticum*** (Latreille, 1798)
Figs. 22–24.

Formica erratica Latreille, 1798:44.

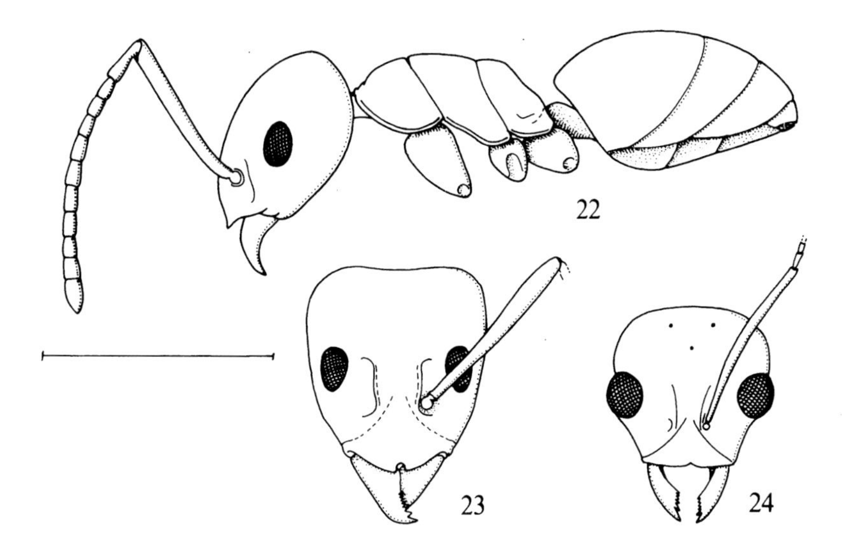

Figs. 22–24. *Tapinoma erraticum* (Latr.). – 22: worker in profile; 23: head of worker in dorsal view; 24: head of male in dorsal view. Scale: 1 mm.

Worker. Dark brown to black; head wedge shaped widening posteriorly; clypeus with median notch as wide as or wider than deep. Dorsum of alitrunk and appendages without standing hairs. Body covered with adpressed white pubescence and very finely punctured. Antennae 12 segmented, palp formula 6,4. Length: 2.6–4.2 mm.

Queen. As worker but with enlarged flattened alitrunk. Length: 4.5–5.5 mm.

Male. Dark brown to black; clypeus notched; scapes long, overreaching occipital border; head wedge shaped; external genitalia prominently exposed. Subgenital plates as wide as or wider than enclosed median area. Length: 3.5–5.0 mm.

Distribution. Present in Scandinavia only on the islands of Gotland and Öland. It occurs on sandy heaths and along the coast in South England and in the Channel Isles. – Range: throughout Central Europe from Spain to the Caucasus and from the mountains of South Italy to North Germany.

Biology. This is a small black ant, agile and aggressive on disturbance. Colonies usually contain several hundred workers and many queens. Nests are constructed under stones or in bare ground in dry sunny places and loose earth solaria for brood incubation are often constructed during the early summer. This species is partly aphidicolous and partly carnivorous. Alatae are developed in June with flights occurring during July.

Tapinoma melanocephalum (Fabricius, 1793)

Formica melanocephala Fabricius, 1793:353.

Worker. Distinctively bicoloured, head brown, alitrunk and gaster pale yellowish with variable brown patches; legs and antennae very pale. Prothorax laterally compressed, broadening anteriorly. Length: 1.5–2 mm.

Queen. Colour as worker. Alitrunk enlarged. Length: 2.5 mm.

Male. Head and dorsum of alitrunk dark, gaster pale with indefinite dark markings, wings and legs very pale. Length: 2.0 mm.

Biology. This is frequently introduced in plant material or produce from the tropics and has occasionally become established in heated or domestic premises in the British Isles. It is a very well known and widely distributed house pest in the tropics where it lives by scavenging.

SUBFAMILY MYRMICINAE LEPELETIER

This is a large subfamily incorporating about 200 genera with a great diversity of form. All have two segmented pedicels and stings. Pupae are not enclosed in cocoons.

Keys to genera of Myrmicinae

Workers

1 Propodeum without spines or teeth, with smoothly rounded postero-dorsal border 2
– Propodeum bispinose or toothed 3
2 (1) Antennae with 10 segments terminating in two-segmented club (Fig. 85) *Diplorhoptrum* Mayr (p. 64)
– Antennae with 12 segments terminating in three-segmented club (Fig. 81) *Monomorium* Mayr (p. 62)
3 (1) Mandibles sickle-shaped, narrowing to pointed apex without teeth (Fig. 108) *Strongylognathus* Mayr (p. 80)
– Mandibles subtriangular with broad masticatory border 4
4 (3) Postpetiole attached to dorsum of first gastral segment (Figs. 88, 89) *Crematogaster* Lund (p. 66)
– Postpetiole attached to anterior face of gaster 5
5 (4) Postpetiole with ventral lobe drawn out into an angular process or spine 6
– Postpetiole without distinct ventral projection 7
6 (5) Mandibles exceptionally broad but without teeth; head massive, quadrangular, finely striate (Figs. 104, 105) .... *Harpagoxenus* Forel (p. 78)
– Mandibles normal with 5 distinct teeth; head not much wider than pronotum; whole body smooth and shining (Figs. 102, 103) .... *Formicoxenus* Mayr (p. 77)
7 (5) Antennae 11 segmented (Fig. 92) .... *Leptothorax* Mayr (in part) (p. 68)
– Antennae 12 segmented 8
8 (7) Head underneath with two strong longitudinal carinae; anterior margin of clypeus bidentate (Fig. 90) .... *Myrmecina* Curtis (p. 67)
– Head without ventral carinae; clypeus with anterior margin entire 9
9 (8) Postero-lateral borders of clypeus raised into a ridge in front of antennal insertions; sting with a triangular lamelliform appendage apicodorsally; pronotum with angled anterolateral corners in European species (Figs. 110, 111) .... *Tetramorium* Mayr (p. 82)
– Clypeal border not raised; sting without a lamelliform appendage; pronotum with rounded anterolateral corners in North European species 10
10 (9) Median portion of clypeus longitudinally bicarinate, prolonged posteriorly between narrowly set frontal carinae; petiole with long anterior peduncle (Fig. 72) .... *Stenamma* Westwood (p. 60)
– Median portion of clypeus broad not bicarinate; petiole with short stout peduncle 11
11 (10) All tibia with pectinate spurs; last three antennal segments distinctly shorter than rest of funiculus .... *Myrmica* Latreille (p. 40)
– Mid and hind tibial spurs absent or simple; last three antennal segments about as long as rest of funiculus 12

12 (11) Dimorphic species with broad-headed, large workers having mandibles with two teeth apically, separated from the basal tooth by a long diastema; minor workers have oval heads with multidentate mandibles (Figs. 76–79)
Pheidole Westwood (p. 61)
– Monomorphic with all workers of even size; mandibles with five teeth in decreasing size from base to apex (Fig. 92) .... *Leptothorax* Mayr (p. 68)

Queens

1 Postpetiole attached to dorsum of first gaster segment
Crematogaster Lund (p. 66)
– Postpetiole attached to anterior face of first gaster segment 2
2 (1) Gaster with a broad longitudinal channel; anterior margin of clypeus with a median notch (Fig. 107) .... *Anergates* Forel (p. 79)
– Gaster without a longitudinal channel; anterior margin of clypeus entire 3
3 (2) Petiole quadrangular, biturberculate above; ventral surface of head with paired longitudinal carinae .... *Myrmecina* Curtis (p. 67)
– Petiole nodal with distinct anterior peduncle. Ventral surface of head without paired longitudinal carinae 4
4 (3) Propodeum smoothly rounded in side view 5
– Propodeum distinctly angulate in side view or with more or less prominent spines or teeth 6
5 (4) Antennae with 10 segments, terminating in abrupt 2 segmented club (Fig. 84)
Diplorhoptrum Mayr (p. 64)
– Antennae with 12 segments, terminating in 3 segmented club (Fig. 83)
Monomorium Mayr (p. 62)
6 (4) Mandibles narrow falcate, curving inwards to sharp pointed apex
Strongylognathus Mayr (p. 80)
– Mandibles subtriangular with broad masticatory border 7
7 (6) Antennae with 11 segments 8
– Antennae with 12 segments 10
8 (7) Postpetiole with long or enlarged ventral tooth like process 9
– Postpetiole with indistinct minute antero-ventral tooth (Fig. 94)
Leptothorax Mayr (in part) (p. 68)
9 (8) Mandibles with distinct teeth; body shining without sculpture; subpostpetiolar process a long and acute spine .... *Formicoxenus* Mayr (p. 77)
– Mandibles edentate; head sculptured; subpostpetiolar process as a blunt tooth .... *Harpagoxenus* Forel (p. 78)
10 (7) Mandibles with a large apical and preapical tooth well separated from indistinct basal tooth; mesoscutum smooth and shining .. *Pheidole* Westwood (p. 61)
– Mandibles with masticatory margins dentate throughout; mesoscutum sculptured 11

11 (10) Posterolateral portion of clypeus raised into a ridge in front of the antennal insertions. Sting with lamelliform appendage .... *Tetramorium* Mayr (p. 82)

– Clypeus not raised into a ridge posteriorly. Sting without lamelliform appendage 12

12 (11) Tibial spurs of middle and hind legs distinctly pectinate *Myrmica* Latreille (p. 40)

– Tibial spurs of middle and hind legs simple or absent 13

13 (12) Postpetiole with a massively developed ventral process (Fig. 69) *Sifolinia* Emery (p. 58)

– Postpetiole without ventral process or tooth 14

14 (13) Petiole with a long anterior peduncle; median portion of clypeus narrow, distinctly bicarinate; body hairs long and acute .... *Stenamma* Westwood (p. 60)

– Petiole with a short thick peduncle; median portion of clypeus broad without distinct carinae; body hairs short and blunt .. *Leptothorax* Mayr (in part) (p. 68)

Males

1 Apterous, always without wings 2

– Alate, wings always present 3

2 (1) Antennae with 10 or 11 segments; appearance pupoidal with dull sculptured integument; anterior margin of clypeus notched (Fig. 106) *Anergates* Forel (p. 79)

– Antennae with 12 segments; appearance worker-like with shining unsculptured integument; anterior margin of clypeus without median notch *Formicoxenus* Mayr (p. 77)

3 (1) Antennae with 10 segments, with elongate second funicular segment 4

– Antennae with 12 or 13 segments 5

4 (3) Mandibles edentate, curved, narrowing to pointed apex (Fig. 109) *Strongylognathus* Mayr (p. 80)

– Mandibles dentate; subtriangular .... *Tetramorium* Mayr (p. 82)

5 (3) Postpetiole attached to dorsum of first gaster segment *Crematogaster* Lund (p. 66)

– Postpetiole attached to anterior face of first gaster segment 6

6 (5) Antennae with 12 segments 7

– Antennae with 13 segments 10

7 (6) Propodeum smoothly rounded; notauli absent (Fig. 87) *Diplorhoptrum* Mayr (p. 64)

– Propodeum biangulate or with distinct teeth or spines; notauli present 8

8 (7) Mandibles with 5 distinct teeth; antennal scape longer than half funiculus *Sifolinia* Emery (p. 58)

– Mandibles reduced, edentate; antennal scapes shorter than half funiculus 9

9 (8) Postpetiole with large ventral tooth-like process ..... *Harpagoxenus* Forel (p. 78)
– Postpetiole with a minute antero-ventral tooth (Fig. 95)
Leptothorax Mayr (in part) (p. 68)
10 (6) Mandibles extremely reduced, non-functional; petiole quadrangular without anterior peduncle; wings very dark (Fig. 91) .... *Myrmecina* Curtis (p. 67)
– Mandibles with 3 or more teeth; petiole nodal with distinct anterior peduncle; wings light 11
11 (10) First funicular segment short and bulbous; ocelli enlarged and protuberant (Fig. 80) .... *Pheidole* Westwood (p. 61)
– First funicular segment elongate; not swollen; ocelli not markedly protuberant 12
12 (11) Notauli and parapsidal furrows absent ....*Monomorium* Mayr (p. 62)
– Distinct Y-shaped notauli present 13
13 (12) Mid and hind tibiae with distinct pectinate spurs (Fig. 3)
Myrmica Latreille (p. 40)
– Mid and hind tibial spurs simple or absent 14
14 (13) Anterior peduncle drawn out and elongate; propodeum bidentate (Fig. 75)
Stenamma Westwood (p. 60)
– Anterior peduncle short and thick; propodeum simply angled or rounded (Fig. 101) .... *Leptothorax* Mayr (in part) (p. 68)

TRIBE MYRMICINI SMITH

Genus *Myrmica* Latreille, 1804

Myrmica Latreille, 1804:175.
Type-species: *Formica rubra* Linné, 1758.

This holarctic genus includes about 70 species of which 10 occur in North Europe. Head oval, clypeus rounded, frontal lobes prominent; alitrunk with pronotum rounded antero-laterally, a well defined mesopropodeal furrow and, in most species, strongly developed propodeal spines. Pedicel with two large nodes – the petiole with a stout antero-ventral tooth and the postpetiole which, rubbed against the fine transverse striae on the first gaster segment, gives an audible stridulation. All tibial spurs are distinctly pectinate except in a very few semi-parasitic species. Gaster in female castes armed with a strong sting. Fore-wing in male and queen have the cubital cell characteristically partially divided by a transverse vein. Palp formula 6, 4; antennae 12 segmented in female caste, 13 in male.

Myrmica ants are robust, deliberate moving, nesting in stumps, under stones or in banks. Colonies are relatively small with individuals numbering from a few hundred to about 5000 according to species.Foraging is on the ground surface.

Keys to species of *Myrmica*

Workers

1 Antennal scape long and slender, gently curved near the base. Frontal triangle entirely smooth and shining 2

– Antennal scape sharply curved near the base or distinctly angled, with or without a toothlike or lamellar extension at the head 3

2 (1) Petiole in profile with large truncate dorsal area, posteriorly with a distinct step down to its junction with the postpetiole. Infra-spinal area transversely striate; petiole nodes rugose; propodeal spines as long as the distance between their tips (Fig. 26) .... 6. *ruginodis* Nylander

– Petiole in profile with dorsal surface a small rounded dome sloping posteriorly, without a distinct step to its junction with the postpetiole. Infra-spinal area smooth; petiole nodes shining, without rugose sculpture. Propodeal spines shorter than the distance between their tips (Fig. 25) .... 5. *rubra* (Linné)

3 (1) Antennal scape abruptly curved near the base but never angled nor with lamellar outgrowth 4

– Antennal scape angled sharply near the base, with or without lamellar extension at bend 6

4 (3) Body sculpture including nodes, clypeus and frontal triangle with coarse longitudinal rugae; petiole massively domed; propodeal spines rather blunt, incurved and subparallel from above (Fig. 27) .... 12. *sulcinodis* Nylander

– Body sculpture finely striate or rugulose; frontal triangle striate or sculptured apically, only with lower part smooth and shining 5

5 (4) Head longer than broad with wide frons; petiole smooth, in profile simply angled without a distinct truncated dorsum; postpetiole cubical (Fig. 29)
7. *rugulosa* Nylander

– Head not longer than broad, frons narrower with diverging lobes. Petiole striated, with a distinct truncated dorsal area; postpetiole higher than long in profile (Fig. 28) .... 3. *gallieni* Bondroit

6 (3) Antennal scape with a distinct transverse upright flange at bend seen as a toothlike projection in profile 7

– Antennal scape simply angled or with a more or less large lateral lamella and a distinct ridge running forward from the bend 8

7 (6) Frons about ¼ head width, with small narrow widely diverging lobes to accommodate the large lamelliform flange. Postpetiole in profile low, only slightly higher than long; petiole with rounded antero-dorsal angle, mesopropodeal furrow shallow (Figs. 31, 40) .... 10. *schencki* Emery

– Frons about ⅓ head width, with more prominent blunter frontal lobes; lamelliform flange variable, often very large appearing as a distinct vertical tooth in profile. Postpetiole distinctly higher than long in profile, petiole with anterior and dorsal faces meeting at a sharp angle, mesopropodeal furrow deep (Figs. 32, 41) .... 4. *lobicornis* Nylander

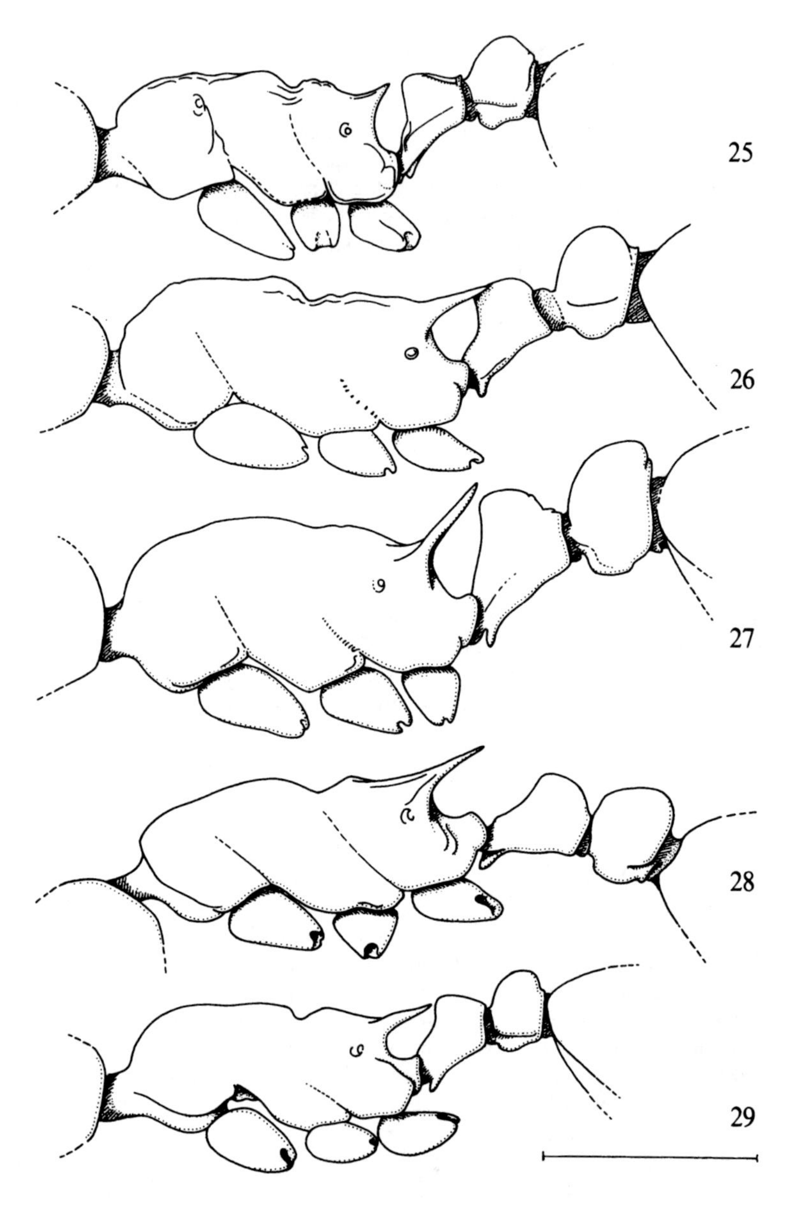

Figs. 25–29. Alitrunk in profile of workers of *Myrmica.* – 25: *rubra* (L.); 26: *ruginodis* Nyl.; 27: *sulcinodis* Nyl.; 28: *gallieni* Bondr.; 29: *rugulosa* Nyl. Scale: 1 mm.

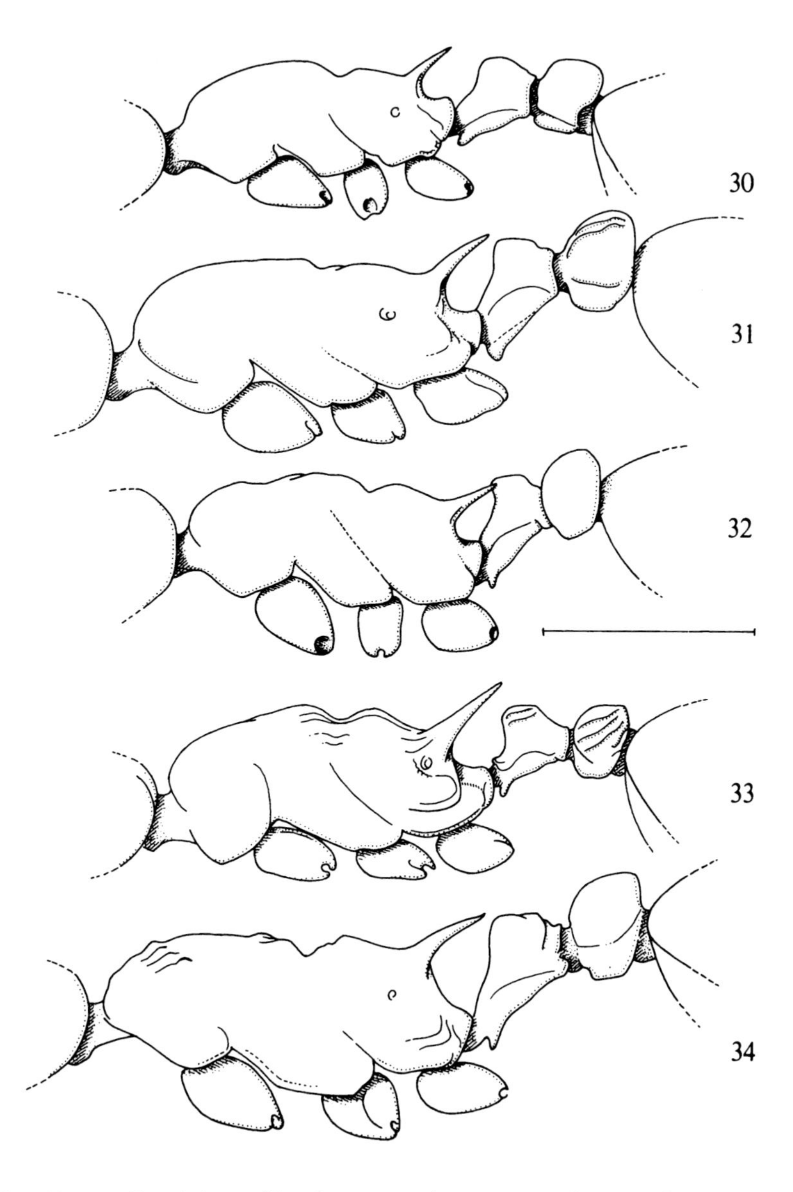

Figs. 30–34. Alitrunk in profile of workers of *Myrmica*. – 30: *specioides* Bondr.; 31: *schencki* Emery; 32: *lobicornis* Nyl.; 33: *scabrinodis* Nyl.; 34: *sabuleti* Mein. Scale: 1 mm.

8 (6) Petiole in profile with dorsal area curving backward to the junction with the postpetiole without a distinct step. Postpetiole about as long as wide from above, slightly higher than long in profile (ratio 10:8) (Figs. 30, 36)
11. *specioides* Bondroit

– Petiole in profile high with a flattened dome or truncate dorsal area, with a distinct posterior edge falling sharply to the junction with the postpetiole. Postpetiole wider than long from above, clearly higher than long in profile (ratio 10:7) 9

9 (8) Antennal scape with massively developed lateral extension at bend; petiole in profile a flattened dome (Figs. 34, 39) .... 8. *sabuleti* Meinert

– Antennal scape simply but sharply angled or with inconspicuous semicircular extension at bend; petiole with distinct truncate dorsal area (Figs. 33, 37)
9. *scabrinodis* Nylander

Queens

1 Antennal scape simply curved near base, without angle or lamellar outgrowth 2

– Antennal scape bent at an angle near base, with or without a lamellar outgrowth at bend 6

2 (1) Antennal scape long and slender, gently curved near the base; frontal triangle entirely smooth and shining 3

– Antennal scape abruptly curved near base; frontal triangle at least in part striated or sculptured 4

3 (2) Propodeal spines long and coarse; infra-spinal area transversely striate; petiole and postpetiole sculptured .... 6. *ruginodis* Nylander

– Propodeal spines short; infra-spinal area smooth; petiole and postpetiole shining without sculpture .... 5. *rubra* (Linné)

4 (2) Frontal triangle completely and coarsely striate; petiole in profile an irregular dome .... 12. *sulcinodis* Nylander

– Frontal triangle in part smooth and shining; petiole either simply angled or clearly truncate in profile 5

5 (4) Postpetiole cubical in profile, not higher than long. Petiole simply angled with very short dorsal area .... 7. *rugulosa* Nylander

– Postpetiole distinctly higher than long in profile. Petiole with distinct truncate dorsal area .... 3. *gallieni* Bondroit

6 (1) Antennal scape with upright lamelliform flange appearing as a vertical tooth in profile 7

– Antennal scape either simply angled or with a lateral extension at bend 8

7 (6) Postpetiole distinctly higher than long in profile; frontal ridges not closely approximated, frontal lobes strongly developed; predominant colour brownish black .... 4. *lobicornis* Nylander

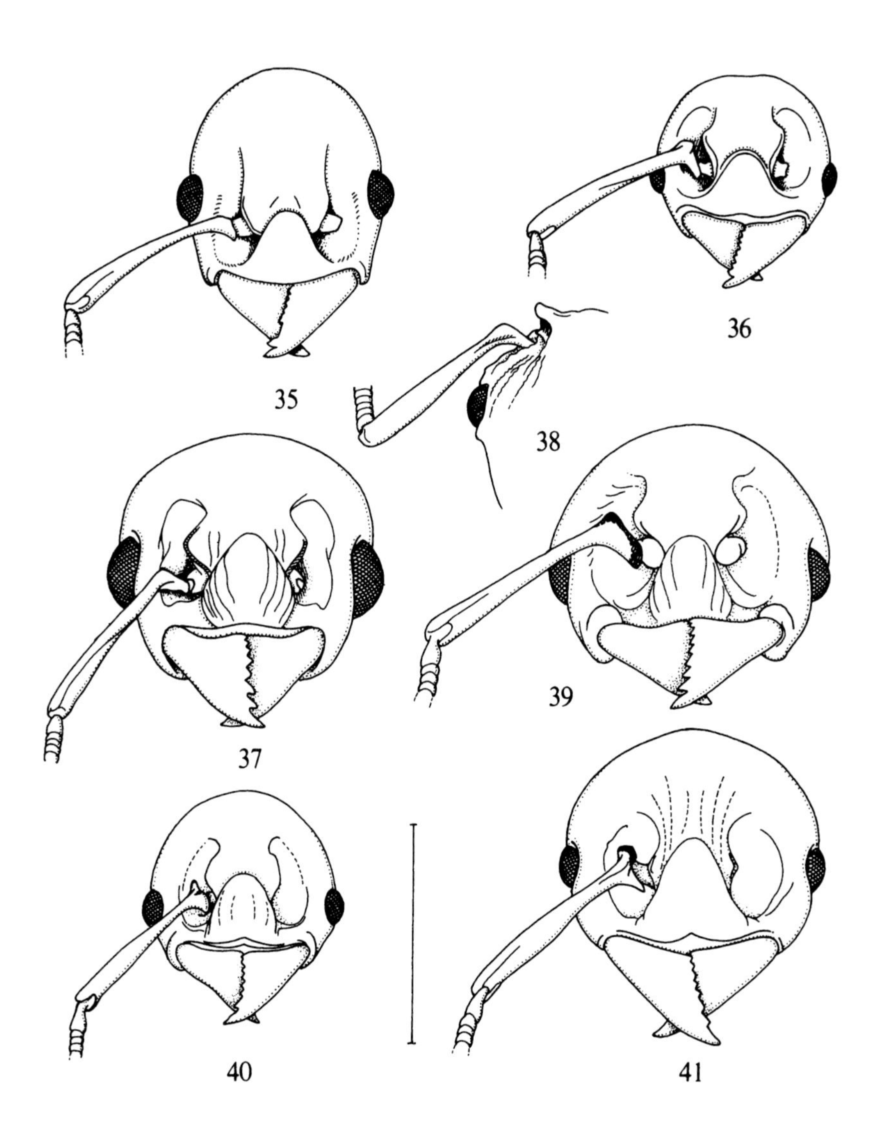

Figs. 35–41. Head of *Myrmica*-workers in antero-dorsal view, except 38, which shows antennal scape insertion in posterior view. – 35: *rugulosa* Nyl.; 36: *specioides* Bondr.; 37: *scabrinodis* Nyl.; 38: same; 39: *sabuleti* Mein.; 40: *schencki* Emery; 41: *lobicornis* Nyl. Scale: 1 mm.

– Postpetiole low in profile; frontal ridges closely approximated with frontal lobes narrow and widely divergent; predominant colour reddish brown 10. *schencki* Emery

8 (6) Petiole in profile with dorsal surface curving evenly into posterior surface without a distinct posterodorsal edge, narrowly rectangular from above 11. *specioides* Bondroit

– Petiole in profile with dorsal surface flattened and a distinct posterodorsal edge, broadly quadrate from above 9

9 (8) Petiole with anterior and dorsal surfaces meeting at a blunt angle. Antennal scape with a lateral projection at bend and a distinct ridge running forward from the bend 10

– Petiole with anterior and dorsal surfaces meeting at a sharp angle. Antennal scape simply angled or with small semi-circular extension without a distinct ridge running forward from the bend .... 9. *scabrinodis* Nylander

10 (9) Head width 1.15 mm or greater , postpetiole narrower than maximum width across frontal laminae .... 8. *sabuleti* Meinert

– Head width less than 1.12 mm; postpetiole wider than maximum width across frontal laminae .... *hirsuta* Elmes

Males

1 Antennal scape long, reaching or overreaching occipital border 2

– Antennal scape short, not reaching beyond ocelli 5

2 (1) Antennae slender with gentle curvature near base; frontal triangle smooth 3

– Antennae sharply bent or curved near base; frontal triangle at least in part sculptured 4

3 (2) Central area of head between frontal ridges longitudinally rugulose. Hind tibiae with long projecting hairs (Fig. 53) .... 5. *rubra* (Linné)

– Central area of head between frontal ridges smooth. Hind tibiae with sparse short hairs (Fig. 52) .... 6. *ruginodis* Nylander

4 (2) Mesoscutum in front of notauli rugose; frontal triangle coarsely striated. Antennal scapes simply curved (Fig. 44) .... 12. *sulcinodis* Nylander

– Mesoscutum in front of notauli smooth; frontal triangle in part smooth. Antennal scape bluntly angled near base (Fig. 45) .... 4. *lobicornis* Nylander

5 (1) Antennal scapes at least at long as first four funicular segments together (Fig. 50) 10

– Antennal scapes not longer than first three funicular segments together 6

6 (5) Body and appendage hairs abundant and long. Longest hairs are longer than maximum width of hind tibiae (Fig. 55) .... 9. *scabrinodis* Nylander

– Appendage hairs not as long as maximum width of hind tibiae 7

7 (6) Second funicular segment more than twice as long as wide; scape angled near base (Fig. 51). Median area of head in front of occellus narrowly depressed 10. *schencki* Emery

- Second funiculus segment less than twice as long as wide; scape not angled near base. Dorsum of head without depression 8

8 (7) Petiole low with anterior and dorsal faces meeting at a gently rounded obtuse angle. Postpetiole in profile longer than high (Fig. 63)
11. *specioides* Bondroit

- Petiole high with anterior and dorsal faces meeting at a right angle. Postpetiole in profile as high or higher than long 9

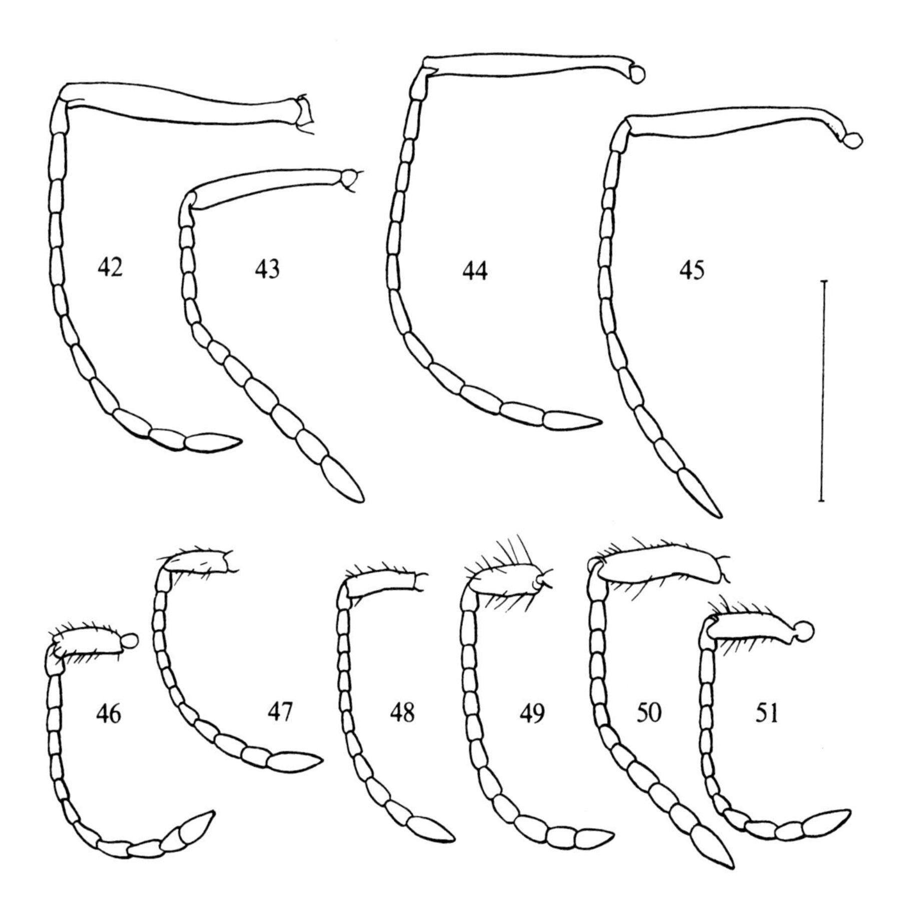

Figs. 42–51. Antenna of *Myrmica*-males. – 42: *ruginodis* Nyl.; 43: *rubra* (L.); 44: *sulcinodis* Nyl.; 45: *lobicornis* Nyl.; 46: *gallieni* Bondr.; 47: *rugulosa* Nyl.; 48: *specioides* Bondr.; 49: *scabrinodis* Nyl.; 50: *sabuleti* Mein.; 51: *schencki* Emery. Scale: 1 mm.

9 (8) Mesoscutum in front of notauli finely sculptured .... 3. *gallieni* Bondroit
– Mesoscutum shining without sculpture (Fig. 3) .... 7. *rugulosa* Nylander
10 (5) Petiole and postpetiole conspicuously hairy with long setae exceeding maximum appendage width (Fig. 68) .... *hirsuta* Elmes
– Petiole and postpetiole with scattered short setae not exceeding maximum appendage width (Fig. 66) .... 8. *sabuleti* Meinert

In the following descriptions indices are given for the worker caste as follows: – Head Index: width immediately in front of eyes × 100 ÷ length from anterior of clypeus to occiput. Frons Index: narrowest width between the frontal ridges × 100 ÷ head width. Frons laminae width: narrowest width between the frontal ridges × 100 ÷ distance across their maximum divergence. These indices are taken from Sadil (1951) as averages of many counts. They provide useful additional criteria for distinguishing the species but because of variability and overlapping measurements between the species, cannot be used for keying them satisfactorily.

3. ***Myrmica gallieni*** Bondroit, 1920

Figs. 28, 46, 65.

Myrmica gallieni Bondroit, 1920:302.
Myrmica rugulosa subsp. *limanica* Arnoldi, 1934:162.
Myrmica jacobsoni Kutter, 1963:133.
Myrmica bergi Ruzsky; Sadil, 1951 (misidentification).

Worker. Light to dark reddish brown. Antennal scapes slender, obliquely curved near the base. Head and alitrunk longitudinally striate; frontal triangle with striae at apex and lower portion smooth, somewhat shining. Propodeal spines thin, space between their bases smooth. Mesopropodeal furrow deep. Petiole with short truncate dorsal area. Postpetiole higher than wide. Head Index: 84.1, Frons Index: 43.8, Frontal Laminae Index: 90.3. Length: 4.5–5.0 mm.

Queen. Dark reddish brown. Sculpture, antennae, propodeal spines and pedicel as worker; distinguished from pale examples of *M. sulcinodis* by the short truncate petiole. Length: 6.0–6.5 mm.

Male. Brownish black. Scape short, straight equal to 3 following segments, second funiculus segment nearly twice as long as wide. Propodeum bluntly tuberculate. Petiole high with a long anterior face, postpetiole distinctly higher than wide in side view. Tarsal hairs on extensor surface longer than those on underside. Length: 5.5–6.0 mm.

Distribution: Very local. Denmark: LFM, Rødby, sand dunes (leg. Collingwood, 1963, 1974). – Sweden: Gotland (H. Lohmander leg. 1934). – Finland: N, Täktom. – Range: France to Western USSR, Czechoslovakia to Baltic, local.

Biology. In general behavior it appears to resemble *M. rubra* (L.) being somewhat aggressive and stinging freely. In the Rødby dunes it was nesting deep in the ground with simple entrance holes. Its apparent decline at this site may be associated with its greater demand for high summer temperatures than that of the commoner N. European species which would tend to displace it. *M. sabuleti* Mein., *M. scabrinodis* Nyl., *M. rubra* (L.) were all present in the immediate vicinity of *M. gallieni* at Rødby in 1974.

Note. This interesting species was found in local abundance by Jacobson (1939) from a wide area of the North Baltic provinces and offshore islands where it was found colonising salt marsh and sand dune areas. Arnoldi (1934) gives similar situations for the Ukrainian salt marshes under the name of *M. rugulosa* subsp. *limanica*. In Czechoslovakia it was collected in the vicinity of Třebon near the lakes Opatovice and Svet and identified later by Sadil (1951) as *M. bergi* Ruzsky, a larger broadheaded species found in S. Russia, Turkestan and Iraq. Kramer (1950) gave a brief redescription of the worker (as *M. gallieni*) from sandy terrain in the Netherlands.

In Denmark and Fennoscandia it was first recognised in 1963 where it was found abundantly in a part of the coastal dune area near Rødbyhavn in Lolland. By 1974 however the species had practically disappeared from the site. There is a series in the Göteborg Natural History Museum in the Hans Lohmander collection taken in July

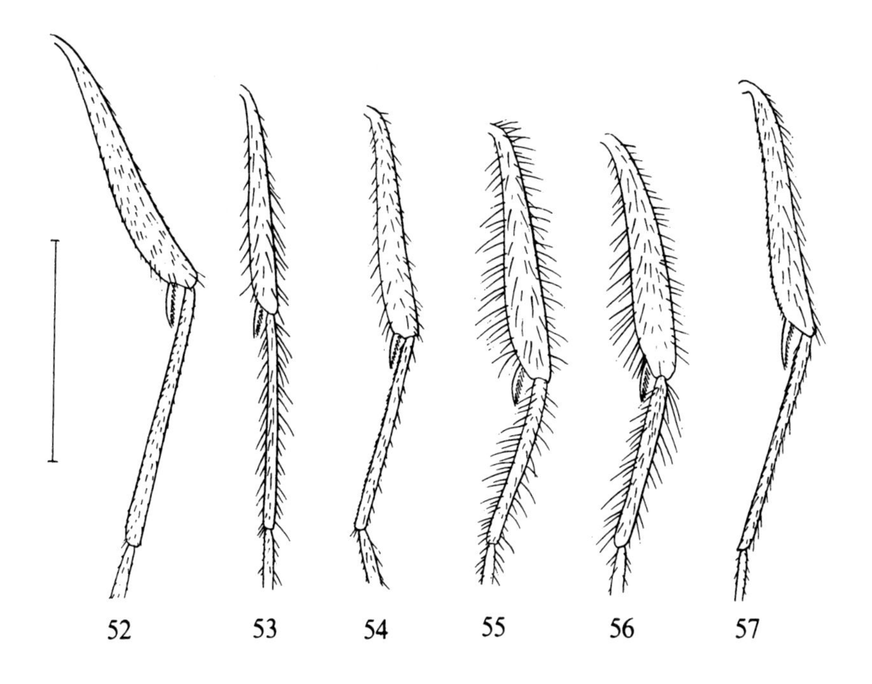

Figs. 52–57. Hind leg in *Myrmica*-males. – 52: *ruginodis* Nyl.; 53: *rubra* (L.); 54: *specioides* Bondr.; 55: *scabrinodis* Nyl.; 56: *sabuleti* Mein.; 57: *schencki* Emery. Scale: 1 mm.

1934 at Horsne on Gotland and it may be presumed that further collecting in coastal areas of S. Sweden and Denmark will reveal more records. There is also a series of workers in the Helsinki Natural History Museum from Täktom on the Hangko peninsula in Nylandia.

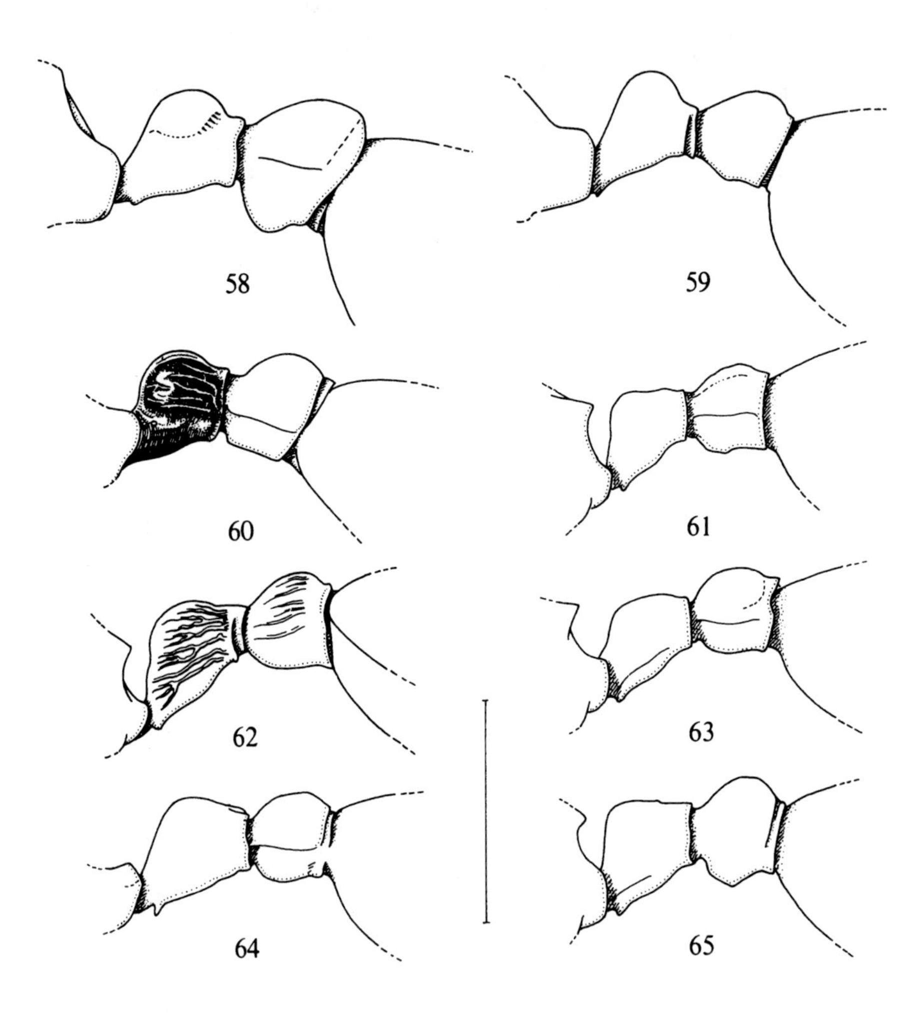

Figs. 58–65. Petiole and postpetiole of *Myrmica*-males. – 58: *ruginodis* Nyl.; 59: *rubra* (L.); 60: *sulcinodis* Nyl.; 61: *rugulosa* Nyl.; 62: *lobicornis* Nyl.; 63: *specioides* Bondr.; 64: *schencki* Emery; 65: *gallieni* Bondr. – Scale: 1 mm.

Myrmica hirsuta Elmes, 1978
Fig. 68.

Myrmica hirsuta Elmes, 1978:131.

Queen. Similar to a microgyne *M. sabuleti* but distinguished by the laterally enlarged postpetiole, wider frons and excessive development of body hairs. Head width: 1.05 mm. Body length: 5.2 mm. Mean postpetiole width: 0.675 mm.

Male. Similar to *M. sabuleti* except in smaller size and more profuse body hairs. Length: 5.3 mm.

Distribution. Rare, Dorset, South England only.

Biology. This species was discovered by Elmes (1978) in a small group of colonies containing apparently normal *M. sabuleti* workers and in some cases normal queens. The small queens were at first assumed to be microgynes of *M. sabuleti* but body pilosity and relative postpetiole measurements were found to be outside the range of that species. In size and appearance *M. hirsuta* resembles *M. myrmecoxena* Forel found once only as a parasite of *M. lobicornis* in Switzerland and also has affinities with the similar but much larger *Myrmica bibikoffi* Kutter (1963).

4. ***Myrmica lobicornis*** Nylander, 1846
Figs. 32, 41, 45, 62.

Myrmica lobicornis Nylander, 1846:932.

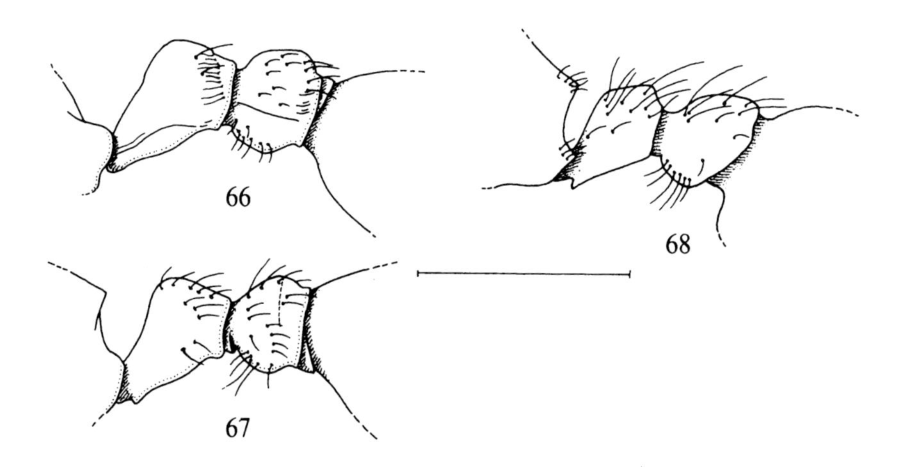

Figs. 66–68. Petiole and postpetiole of *Myrmica*-males. – 66: *sabuleti* Mein.; 67: *scabrinodis* Nyl.; 68: *hirsuta* Elmes. Scale: 1 mm. (68 redrawn from Elmes, 1978).

Worker. Bicoloured reddish brown with head and gaster characteristically darker. Upright tooth-like process at the bend of the antennal scape, frequently very large in Scandinavian samples but very variable in size over its whole geographic range. Frons about $^1/_3$ head width. Petiole high with anterior and dorsal surfaces meeting at a right angle. Postpetiole broadly oval from above. Head Index: 87.8; Frons Index: 30.8; Frontal Laminae Index: 65.5. Length: 4.0–5.0 mm.

Queen. As worker, with head and scutum normally darker. Length: 5.0–5.5 mm.

Male. Black, legs and articulations paler. Antennal scape as long as 5–6 following segments, angled near base. Length: 5.0–5.5 mm.

Distribution. Locally common throughout Denmark, Fennoscandia and the British Isles but excluding Ireland. – Range: Portugal to Central Russia, Appenines to Arctic Norway.

Biology. It is a mountain species in Central and S. Europe but in the north occurs equally on lowland heath and in open woodland. Although widely distributed it is not abundant and occurs in isolated single queen colonies nesting in peat or under stones. It is commonly found as single foraging workers and is one of the least aggressive members of the genus.

5. **Myrmica rubra** (Linné, 1758)
Figs. 25, 43, 53, 59.

Formica rubra Linné, 1758:580.
Myrmica laevinodis Nylander, 1846a:927.
Myrmica rubra (Linné); Yarrow, 1955b:113.

Workers. Yellowish brown. Sculpture dilute; frontal triangle and subspinal areas smooth and shining. Antennal scapes long and slender. Petiole node with short indistinct dorsal area sloping evenly without definite break to its junction with the postpetiole. Head Index: 79.5; Frons Index: 49.4; Frontal Laminae Index: 92.7. Length: 3.5–5.0 mm.

Queen. As worker. Length: 5.5–7.0 mm (microgynes 4.5–5.5 mm).

Male. Body colour dark with appendages lighter. Head rugose but rest of sculpture dilute with petiole, postpetiole, area between notauli and also frontal triangle smooth and shining. The funiculus segments are more slender and shorter than in *M. ruginodis* Nyl., the scapes are long and slender, obliquely and evenly curved near the base. The tibiae and tarsi have long projecting hairs which provide the easiest distinction from *M. ruginodis*. Length: 4.5–5.5 mm.

Distribution. Locally common throughout Denmark, South and Central Fennoscandia and the British Isles. Found also in the north in warm sheltered areas only (Lofote Islands, Narvik, Oulu). – Range: Portugal and Ireland to E. Siberia, Italy to North Scandinavia.

Biology. This is a lowland species often abundant where it occurs in sheltered valleys, usually in alluvial soil by riversides and on the coast. Colonies are normally polygynous with several to many queens and up to 1000 or more workers (Elmes, 1973b) nesting in the ground or under stones. Microgynes are quite frequent with this species (Collingwood, 1958; Elmes, 1973a). This is the most aggressive of the *Myrmica* species and stings freely. This ant tends aphids more consistently than other members of the genus and is frequently found collecting nectar on the inflorescence of umbelliflorae and other herbs. Mating flights occur in August and are orientated towards high buildings.

6. ***Myrmica ruginodis*** Nylander, 1846
Figs. 7, 26, 42, 52, 58.

Myrmica ruginodis Nylander, 1846:929.

Worker. Pale to dark reddish brown. Propodeal spines long and robust; area between their bases laterally striate, frontal triangle smooth and shining. Head and alitrunk coarsely longitudinally rugulose. Antennal scapes long and slender, gently and obliquely curved near their base. Petiole in profile massive with truncate dorsal area and abrupt step posteriorly to its junction with the postpetiole gives the easiest distinction from *M. rubra* (L.). Head Index: 77.5; Frons Index: 48.3; Frontal Laminae Index: 91.3. Length: 4.0–6.0 mm.

Queen. As worker. Length: 5.5–7.0 mm (microgynes 4.5–5.5 mm).

Male. Large and robust, characteristically paler than *M. rubra,* with long slender antennal scapes and clearly elongate funiculus segments. The frontal triangle and space between the frontal ridges are smooth and shining. Tibial and tarsal hairs are short, sparse and subdecumbent. Length: 5.0–6.0 mm.

Distribution. Common throughout Denmark and all Fennoscandia and all British regional areas. – Range: throughout Northern Eurasia to Japan.

Biology. This common species is abundant throughout the woodlands and high moorlands of North Europe to the North Cape. Brian and Brian (1949) showed that this species occurred in two incompletely dimorphic races, one polygynous with many small queens approaching the microgyne condition and one monogynous with single large queens which they termed var. *microgyna* and var. *macrogyna* respectively; *microgyna* was found to readily accept strange queens and to occur in more stable habitats often forming groups of nests as is common with *M. rubra; macrogyna* is more aggressive and hostile to strange queens, has more populous nests and is more generally distributed, predominating in woodland and more transitory habitats (Brian and Brian, 1955). Both forms occur in Scandinavia but cannot in conventional taxonomy be regarded as either distinct species or biotopic subspecies because of wide overlap in morphology and habitat. Mating flights occur in August near or on the ground.

7. ***Myrmica rugulosa*** Nylander, 1849
Figs. 3, 29, 35, 47, 61.

Myrmica rugulosa Nylander, 1849:32.

Worker. Pale reddish yellow with dilute sculpture. Head distinctly longer than broad; petiole narrowly rectangular from above, simply angled in side view with very short truncate dorsal area. Antennal scapes sharply but evenly bent near base, without trace of an angular projection or process. Frontal triangle mostly smooth but partly striate at apex in Fennoscandian samples. Head Index: 80.2; Frons Index: 52.7; Frontal Laminae Index: 95.4. Length: 3.0–4.3 mm.

Queen. As worker. Length: 4.8–5.2 mm.

Male. With short antennal scapes; the area between the notauli is smooth and shining without trace of sculpture. In profile the anterior and posterior faces of the petiole meet at a near right angle without a dorsal dome or truncation. The hairs on the extensor surface of the hind tarsi are distinctly longer than those on the underside. Length: 4.0–4.8 mm.

Distribution. Local in South and Central Sweden, Denmark and South Finland. Not recorded from Norway or British Isles. – Range: Central Europe from Pyrenees to Siberia and Italy to Central Sweden.

This is a small slender species, in Scandinavia found locally nesting in lowland sandy heath or open pasture frequently near the coast, but more generally distributed in Europe ascending to 1000 m in the Central Pyrenees. The nests are deep in the ground with circular crater like openings to the outside. The species is polygynous and a single colony may contain many thousands of workers. These tend to forage in files unlike most other *Myrmica* species which forage singly.

8. ***Myrmica sabuleti*** Meinert, 1861
Figs. 14, 34, 39, 50, 56, 66.

Myrmica sabuleti Meinert, 1861:327.

Worker. Reddish brown. Antennal scapes are sharply angulate with a longitudinal keel running forward from the bend and a more or less massive lateral extension, which in Scandinavian samples is frequently curved up to appear as a large semiupright tooth seen from behind. The petiole node is more rounded and usually less truncate than in *M. scabrinodis* Nyl. and the epinotal spines are relatively longer but these features are too variable for certain discrimination between the species in all cases. Head Index: 85.6; Frons Index: 36.8; Frontal Lamina Index: 66.5. Length: 4.0–5.0 mm.

Queen. As worker. Length: 5.5–6.5 mm.

Male. Large and robust, with the antennal scape equal in length to between 4 and 5 following funiculus segments. Appendage hairs are shorter than in *M. scabrinodis* and on the antennae do not exceed their appendage width. Length: 5.0–6.0 mm.

Distribution. Local in Denmark, South and Central Sweden, South Norway and South Finland. Locally common in the British Isles including Scotland and Ireland. - Range: South Europe to Central Scandinavia, Portugal to Urals.

Biology. This is a robust species usually nesting in sun exposed sheltered sites, often in groups of small nests each containing up to a 1000 or more workers with a few queens. It is characteristically larger and more brightly coloured than the similar *M. scabrinodis* and easy to distinguish in Scandinavia where the scape development is relatively massive equivalent to the form described as var. *lonae* Finzi. Nests are usually located under stones but unlike *M. scabrinodis* seldom or never in tree stumps or in boggy land.

9. ***Myrmica scabrinodis*** Nylander, 1846
Figs. 14, 33, 37, 49, 55, 67.

Myrmica scabrinodis Nylander, 1846:930.
Myrmica pilosiscapus Bondroit, 1920:301.
Myrmica rugulosoides Forel, 1915:29.

Worker: Yellow red to blackish brown according to habitat. The antennal scape is sharply angled and sinuate near the base, often with a slight lateral extension in the larger more deeply sculptured forms. The petiole has a distinctly concave anterior face which meets the truncate dorsal surface at a sharp angle. Head Index: 85.6; Frons Index: 36.8; Frontal Laminae Index: 66.5. Length: 4.0–5.0 mm.

Queens. As worker. Length: 5.5–6.5 mm.

Male. Brown to brownish black with profuse long outstanding body and appendage hairs distinctly longer than their appendage width; the antennal scape is short and stout, not longer than the three following funiculus segments. Length: 5.0–6.0 mm.

Distribution. Common throughout Denmark, Fennoscandia and the British Isles. - Range: Throughout Europe.

Biology. The species has variable habits, being found in a very wide range of habitats. In southern areas it is often associated with the meadow ant *Lasius flavus* (Fabr.) living in part of the mound nest and preying on the *L. flavus* workers but may be equally common in woodland, coastal sand, gravel river banks, peat bogs and moorland. Individual nests are small, situated under stones, in tree stumps or in the ground with a few hundred workers and one or a few queens. The alatae fly in August, pairing occurring in the air.

Note. This species is very variable in size and colour. The smaller samples can be confused with *M. rugulosa* or *M. specioides* but can be distinguished by the sharply truncate petiole and narrower frons. In Northern Britain colonies of very dark, often deeply sculptured workers are frequently seen with the scape having a slight semi-circular extension at the bend. This is equivalent to the form described as *M. pilosiscapus* Bondroit

(1920) and redescribed by Sadil (1951), but there is no clear difference in male or female castes between this and the accepted *M. scabrinodis* for a specific distinction. Similarly *M. rugulosoides* Forel (1915) was described as small and weakly sculptured as is frequent in many colonies of this variable species. In some nests large macrergate workers occur with deeper colour and sculpture among the smaller lighter coloured workers while the males of all these variable forms have the constant specific features of long body and appendage hairs and short thick scape.

10. ***Myrmica schencki*** Emery, 1895
Figs. 31, 40, 51, 64.

Myrmica rubra scabrinodis var. *schencki* Emery, 1895:315.

Worker. Brownish red with gaster and sometimes head darker. Frontal triangle striate. Antennal scape sharply angled near base, with an upright flange at the bend fitted closely into the thin divergent frontal ridge. Frons very narrow, about ¼ head width. Mesopropodeal furrow shallow and postpetiole low, somewhat cubical in profile and spherical from above. Head Index: 88.4; Frons Index: 24.5; Frontal Laminae Index: 63.3. Length: 4.0–5.5 mm.

Queen. As worker. Length: 5.0–6.0 mm.

Male. Scapes very short, angled; second funiculus segment elongate. Appendage hairs short, subdecumbent. Petiole long and low, often striate at dorsolateral margins. Length: 4.0–4.5 mm.

Distribution. Local in Denmark and southern areas of Finland, Sweden and Norway, also S. England, Wales and Ireland. – Range: South Europe to South Scandinavia.

Biology. This is an interesting species somewhat resembling a large paler *M. lobicornis* but distinguished by the lower more cubical postpetiole. According to the Danish myrmecologist Chr. Skøtt, this species differs from other European *Myrmica* in having no winter brood, is mainly nocturnal and derives much of its food from the glandular excretions of low herbage such as *Hypochaeris* and *Hieracium* spp. The entrance to the nest is frequently built up as a collar of vegetable detritus (Bisgaard, 1944). Colonies are single queened and isolated, situated in sandy banks and dry pasture. Alatae are found in August, mating occurring on the ground near the nest.

11. ***Myrmica specioides*** Bondroit, 1918
Figs. 30, 36, 48, 54.

Myrmica specioides Bondroit, 1918:100.
Myrmica scabrinodis ssp. *rugulosa* var. *rugulosoides* Forel, 1915: 29.
Myrmica rugulosoides var. *striata* Finzi, 1926:117.
Myrmica puerilis Stärcke, 1942:24.
Myrmica balcanica Sadil, 1951:253.
Myrmica balcanica var. *scabrinoides* Sadil, 1951:255.

Worker. Yellow red to reddish brown. Antennal scapes sharply angulate at bend with a more or less distinct lateral expansion. Petiole narrow, rectangular from above, in side view sloping evenly from the anterodorsal crest to its junction with the postpetiole. Postpetiole spherical almost cubical in side view, only slightly higher than wide. Head Index: 84.6; Frons Index: 40.6; Frontal Laminae Index: 78.3. Length: 3.0–4.5 mm.

Queen. As worker. Length: 5.0–5.5 mm.

Male. With short scape equal in length to three following segments; appendages more slender than in *M. scabrinodis* with hairs not longer than their appendage width, those on the underside of the hind tarsae being distinctly shorter than those on the extensor surface; petiole low with shallow dorsal curvature. Length: 5.0 mm.

Distribution. Rare. Denmark: LFM, Rødby, sand dunes, 1 colony Sept. 1974. – Finland: N, Täktom; Ab, Rymätyllä. – England: East Kent only. – Range: Spain to W. Russia; Italy to South Finland.

Biology. This is a rather local species in Europe but likely to be overlooked through confusion with *M. scabrinodis* in the female castes and may well occur in other areas of southern Fennoscandia. It is a more slender species with a broader frons, narrow petiole and more spherical postpetiole. The male resembles that of *M. rugulosa* but has the petiole longer and lower with a much flattened dorsal area. In England and Denmark nests occurred in coastal sand and gravel banks with a simple entrance hole. Workers behave more aggressively than *M. scabrinodis* and sting freely. Alatae have been found in August and September.

12. ***Myrmica sulcinodis*** Nylander, 1846
Figs. 27, 49, 60.

Myrmica sulcinodis Nylander, 1846:934.

Worker. Deep reddish with head and gaster darker. Strongly longitudinally rugulose, frontal triangle longitudinally striate. Antennal scapes sharply but evenly curved near base. Petiole high with long anterior face and rounded steeply sloped dorsal area, never truncate. Propodeal spines stout and blunt, curved so that they lie subparallel from above, not divergent. Mesopropodeal furrow shallow. Head Index: 84.7; Frons Index: 42.8; Frontal Laminae Index: 91.4. Length: 4.0–6.0 mm.

Queen. As worker. Length: 5.5–6.8 mm.

Male. Black; frontal triangle and anterior of mesoscutum between notauli striate or rugulose. Length: 5.5–6.5 mm.

Distribution. A common species of upland moors in Scandinavia and Britain, also more locally on lowland heath. – Range: Portugal to East Siberia, Appenines to Arctic Scandinavia.

Biology. This is a characteristic species of relatively well drained heather moorland. It is easily recognised by its generally dark colour with deep red sometimes infuscated alitrunk and legs and its strong sculpture. In Scandinavia it can only be confused with the very local lighter coloured *M. gallieni* with its much deeper mesopropodeal furrow and clearly truncate petiole or with dark forms of *M. ruginodis* which commonly occur on high moorland but always have the frontal triangle smooth and shining and the propodeal spines sharper and more divergent from above. *M. sulcinodis* nests in small colonies of up to 500 workers with single queens in dry peat or sand among heather or under flat stones, in wetter areas occasionally building small mounds of vegetable fragments for brood incubation. This is a strong robust species living by predation and scavenging. The alatae fly in August mating in the air over high ground.

Genus *Sifolinia* Emery, 1907

Sifolinia Emery, 1907:49.
Type-species: *Sifolinia laurae* Emery, 1907.

Queen. General shape as in *Myrmica* but antennal scape relatively thick; mid and hind tibiae with pectinate spurs reduced or absent. Both petiole and postpetiole are produced ventrally into blunt projections.

Male. General shape as in *Myrmica* but lacking mid and hind tibial spurs. Antennae 12 segmented. Postpetiole with a blunt ventral projection, petiole thickened and swollen without defined projection. Notauli with posterior portion of Y tail reduced or absent. Wing venation of *Myrmica* type but variable and often reduced.

This is a genus of workerless ants parasitic on *Myrmica* species, with 4 or 5 species known from the palaearctic region only.

13. ***Sifolinia karavajevi*** (Arnoldi, 1930)
Figs. 69–71.

Symbiomyrma karavajevi Arnoldi, 1930:267.
Sifolinia laurae Emery; Yarrow, 1968 (misidentification).
Sifolinia karavajevi (Arnoldi); Kutter, 1973:258 (redescription).

Queen. Pale yellowish brown to brownish, appendages pale; antennae with long scape slightly and evenly curved near base, 12 segmented with indeterminate 3–4 segmented club, ultimate funiculus segment × 1½ length of penultimate. Propodeal spines strong and blunt, about as long as space between their tips. Occiput in full dorsal view feebly concave or straight, eyes prominent, ocelli distinct, frons broad. Postpetiole developed ventrally as a blunt forward projecting tooth. Head and alitrunk with shallow rugulose striae and scattered punctures. First gaster tergite with very short scattered decumbent hairs only; head, alitrunk and appendages with suberect hairs which are longer and thicker on antennae. Length: 3.2–3.6 mm.

Male. Pale brown; antennae 12 segmented, scape about as long as 7 following segments, slightly curved near base. Postpetiole with blunt ventral projection; notauli V shaped; propodeum bidentate. Head and mesonotum very shining. Scattered hairs over dorsum of gaster, longer and thicker on head, alitrunk and appendages. Length: 3.5 mm.

Distribution. Sweden: Sk., Krankesjön (Douwes, 1977). – Norway: HE, Eidskog (Collingwood, 1976). – Finland: Sa, Ryistiina (leg. Forsslund). – England: Dorset, Hampshire, Surrey, very rare. – Range: very local. S. England to Ukraine, Czechoslovakia to Finland.

Biology. This ant has been recorded sometimes in large numbers and sometimes as one or two individuals in nests of various *Myrmica* host species including *M. rugulosa, M. scabrinodis* and *M. sabuleti.* A colony in Dorset, England, was observed for over 4 years during which time alate queens and males of the parasite were present each season together with workers and worker brood of the host, indicating that egg laying queens of both parasite and *Myrmica* host were surviving together in the same nest. In Norway 2 dealate queens were caught in pitfall traps in July 1974 suggesting that after mating, fertilised queens wander over the ground in search of a colony of the host species.

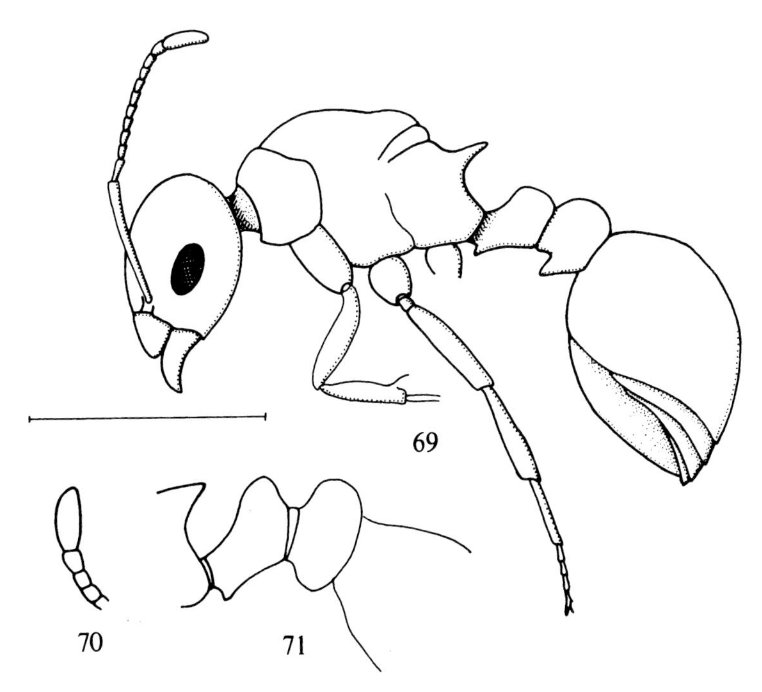

Figs. 69–71. *Sifolinia karavajevi* (Arnoldi). – 69: queen in profile; 70: antennal club of male; 71: petiole and postpetiole of male. Scale: 1 mm.

TRIBE PHEIDOLINI EMERY

Genus *Stenamma* Westwood, 1840

Stenamma Westwood, 1840:83.
Type-species: *Stenamma westwoodii* Westwood, 1840.

Worker and queen. Head slightly longer than broad; antennae 12 segmented, scape not reaching as far as occiput; funicular segments 2–10 not longer than broad; palp formula 4:3; frontal ridges narrowly set about a third head width; clypeus bicarinate; eyes conspicuously small. Anterior peduncle conspicuously elongate.

Male. Antennae 13 segmented, scape about as long as three or four following segments which are all elongate. Eyes enlarged and set anterior to midline. Mesoscutum with Y shaped notauli. Propodeum armed with two short but distinct teeth. Forewings with one cubital and one discoidal cell. The taxonomy of some European species has been investigated by Kutter (1974).

This genus includes about 50 species distributed through the Neotropic, Nearctic and Palaeactic regions with several species in S. Europe. Only one occurs in North and Central Europe.

14. ***Stenamma westwoodii*** Westwood, 1840
Figs. 4, 72–75.

Stenamma westwoodii Westwood, 1840:83.

Worker. Pale to dark rusty red; petiole a simple node with rounded dorsal area, propodeal spines short. Head longitudinally striate in front, alitrunk and back of head with weak reticulate sculpture. Body with erect scattered pale hairs, sparse and decumbent on appendages. Length: 3.5–4.0 mm.

Queen. As worker, only slightly larger; wings pale yellowish. Length: 4.2–4.8 mm.

Male. Blackish brown with mandibles and appendages lighter; head longitudinally striate, dull; mandibles with 3 teeth; wings as queen. Propodeal spines well developed. Length: 3.8–4.2 mm. (Kutter, 1974, draws attention to the original description from English material where the male mandible is said to have 5 teeth whereas all recent specimens in the British Isles have 3 toothed mandibles).

Distribution. Found very locally in Denmark: SZ; Sweden: Sk., Bl., Sm., Gtl. and Vg.; Norway: VA and HO. – Also very locally occurring throughout Southern England to the Midlands, Wales and Southern Ireland. – Range: South and Central Europe from Spain to Caucasus and Italy to South Scandinavia.

Biology. This is an unobtrusive species often taken as solitary workers in woodland. Nests consists of up to 150 workers with a single queen. They may be found in dry well drained woodland under deep stones or among tree roots and under moss. Workers forage during early morning or on dull warm days. This species is partly scavenging and partly predatory on small insects and mites but is slow moving and non-aggressive towards other ant species. Alatae are found in the nests from August to late autumn and have been taken on the wing during September and October.

Genus *Pheidole* Westwood, 1841

Pheidole Westwood, 1841:87.
Type-species: *Atta providens* Sykes, 1835.
Figs. 76–80.

Workers. Markedly dimorphic with large broad headed workers contrasting with smaller workers with narrow oval heads. Mandibles broad with very rounded exterior margins and teeth reduced to 2 or 4 in major workers, multidentate in minor workers. Mesopropodeal suture very deep, mesonotum often prominently raised. Antennae terminating in distinct three segmented club as long as or longer than rest of funiculus.

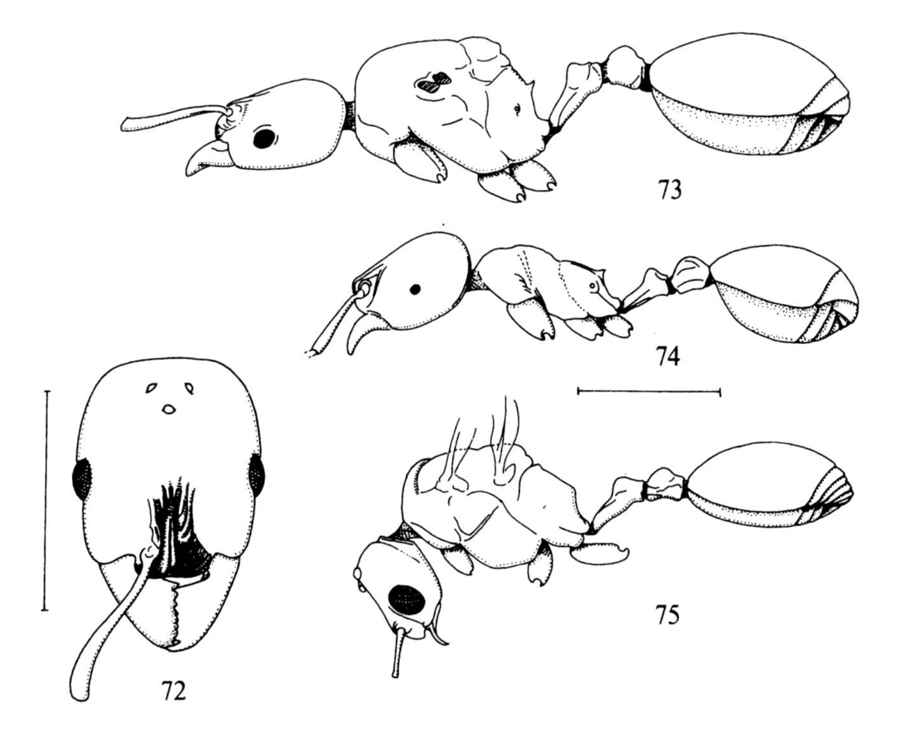

Figs. 72–75. *Stenamma westwoodii* Westwood. – 72: head of worker in dorsal view; 73: queen in profile; 74: worker in profile; 75: male in profile. Scale: 1 mm.

Queen. Head rectangular with well developed eyes and ocelli. Gaster more or less truncated at base.

Male. Antennal scape very short; first funicular segment swollen. Eyes and ocelli enormous, the latter developed on a prominence.

This genus includes many hundreds of similar species occurring throughout the tropics and subtropics with one European species *P. pallidula* (Nylander). Several species may be introduced with imported fruit or on plant material; the commonest cosmopolitan species is *P. megacephala* (Fabricius) and this species and others of the genus have been recorded from time to time from heated premises in England and Denmark.

TRIBE SOLENOPSIDINI FOREL

[emend Ettershank, 1966]

Genus *Monomorium* Mayr, 1855

Monomorium Mayr, 1855:452.
Type-species: *Monomorium minutum* Mayr, 1855.

Worker. Head longer than broad, clypeus bicarinate. Palp formula 1:2 or 2:2. Antennae 11 or 12 segmented with distinct 3 segmented club as long as rest of funiculus – intermediate segments transverse. Mesopropodeal suture deeply marked; propodeum smoothly rounded, unarmed. Gaster basally emarginate with distinct anterior angles.

Queen. Head as in worker. Mesonotum long, overhanging pronotum. Wings with 1 cubital cell and no discoidal cell.

Male. Head in front of ocelli flat and broad; mandibles dentate. Antennae 13 segmented without distinct club; scape not as long as first 3 funicular segments. Mesonotum high, arched without notauli.

This is a worldwide genus with several hundred species mainly occurring in the tropics. There are several cosmopolitan tramp species one of which is well established in North Europe.

Monomorium pharaonis (Linné, 1758)
Figs. 81–83.

Formica pharaonis Linné, 1758:580.

Figs. 76–80. *Pheidole megacephala* (F.). – 76: major worker in profile; 77: minor worker in profile; 78: head of major worker in dorsal view; 79: head of minor worker in dorsal view; 80: head of male in dorsal view. Scale. 1 mm.

Figs. 81–83. *Monomorium pharaonis* (L.). – 81: worker in profile; 82: head of worker in dorsal view; 83: queen in profile. Scale: 1 mm.

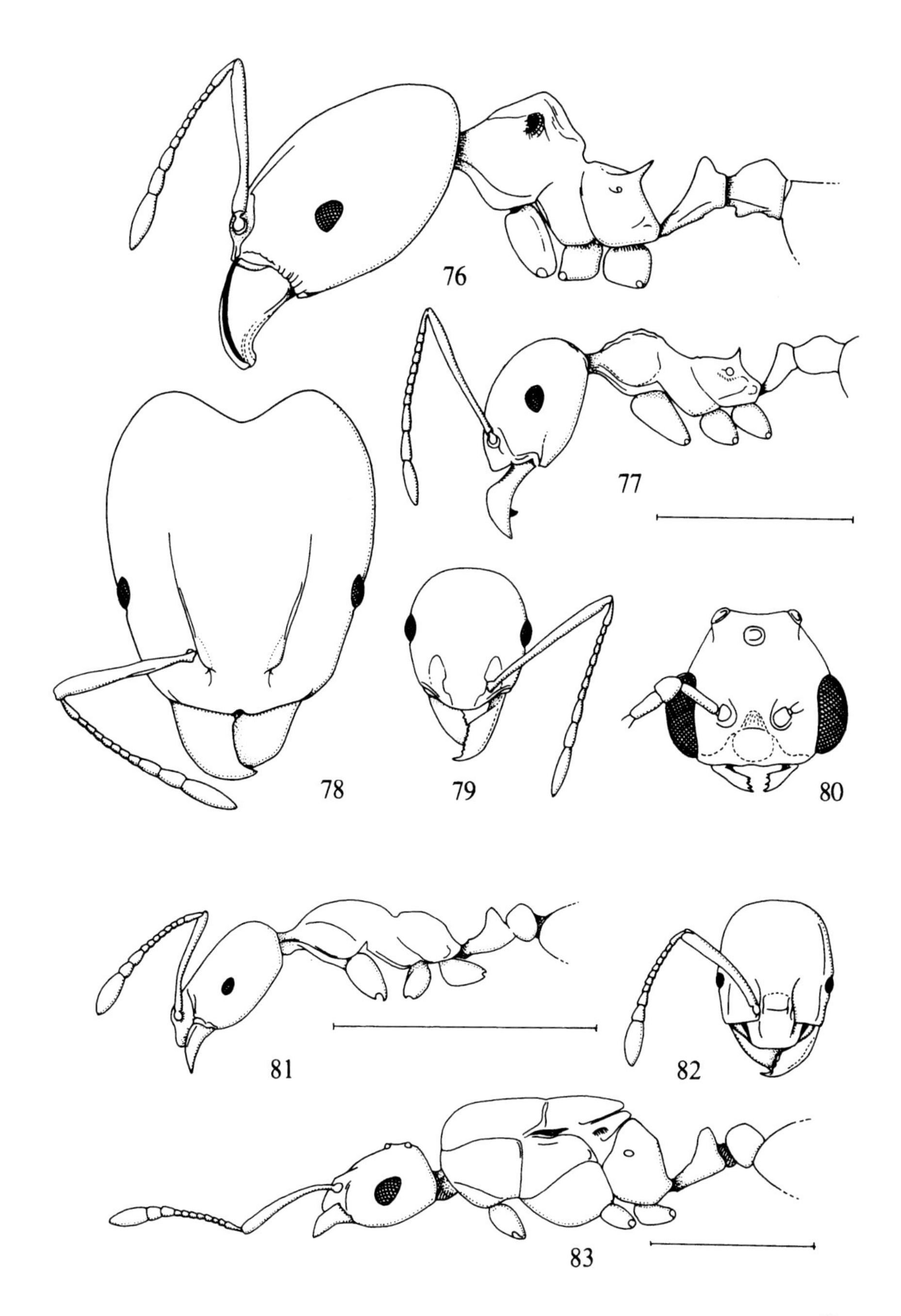
76
77
78
79
80
81
82
83

Worker. Reddish yellow, head and alitrunk closely punctured, dull. Length 2–2.4 mm.

Queen. As worker but with larger eyes and enlarged alitrunk; mesonotum with narrow patch and gaster distinctly darkened posteriorly. Length: 4–4.8 mm.

Male. Black with yellowish appendages, femora and scapes darker. Wings transparent. Eyes very large. Head and alitrunk closely punctured and dull. Length: 3 mm.

Distribution. This is a cosmopolitan species spread by commerce all over the world. In North Europe it is frequently established in heated premises including bakehouses, laundries and hospitals. It has occurred in many places in Denmark, Sweden and Finland and is common in the British Isles where it has been recorded since 1828.

Biology. Colonies are very large, polygynous and polycalic often with several millions of individuals. Workers and queens forage in long trails and live by scavenging on food materials, dead animals and insects. Nests are often sited deep in foundations and are very difficult to eradicate by fumigation or insecticides.

Genus *Diplorhoptrum* Mayr, 1855

[emend Baroni Urbani, 1968]

Diplorhoptrum Mayr, 1855:449.

Type-species: *Formica fugax* Latreille, 1798.

Worker. Small, pale to dark yellow; antennae 10 segmented with distinctly enlarged 2 segmented club. Second funicular segment not longer than broad. Eyes minute. Palp formula 2:2. Clypeus bicarinate with projecting teeth. Propodeum rounded unarmed, mesopropodeal furrow distinct, gaster oval with first segment larger than rest of gaster.

Queen. As worker but larger and darker with relatively massive alitrunk; eyes large and ocelli distinct. Fore wings with 1 cubital cell and 1 discoidal cell, radial cell open.

Male. Mandibles 3 toothed; antennae 12 segmented with first funicular segment swollen and globular. Propodeum unarmed. Volsella with broad band of minute pointed scales.

15. ***Diplorhoptrum fugax*** (Latreille, 1798)
Figs. 84–87.

Formica fugax Latreille, 1798:46.

Worker. Yellow to brownish yellow; sides of head slightly curved. Body and appendages with numerous hairs, head and alitrunk distinctly punctulate but shining. Projecting clypeal teeth short but distinct. Length 1.5–3 mm.

Queen. Blackish brown with punctulate sculpture, very shining. Body and appendages very hairy. Wings slightly fuscous. Length 6–6.5 mm.

Male. Black shining with finely rugose sculpture. Length: 4–4.8 mm.

Distribution. Rare, only recorded from Sk., Öl. and Gtl. in Sweden. - Recorded mainly along the coast in South England from Essex to Cornwall. - Range: Spain to Urals, Italy to Sweden.

Biology. This species lives in populous colonies often deep in the ground or under large stones and is seldom seen above ground. It is often associated with larger *Formica* and *Lasius* species predating on their brood, but nests also occur in isolation. It is mainly predatory and carnivorous but has also been recorded attending root aphids. This ant is aggressive and despite its small size will attack other ants and sting fiercely. Queens and males are very large relative to the workers and are found in August and September with flights occurring on warm days in September.

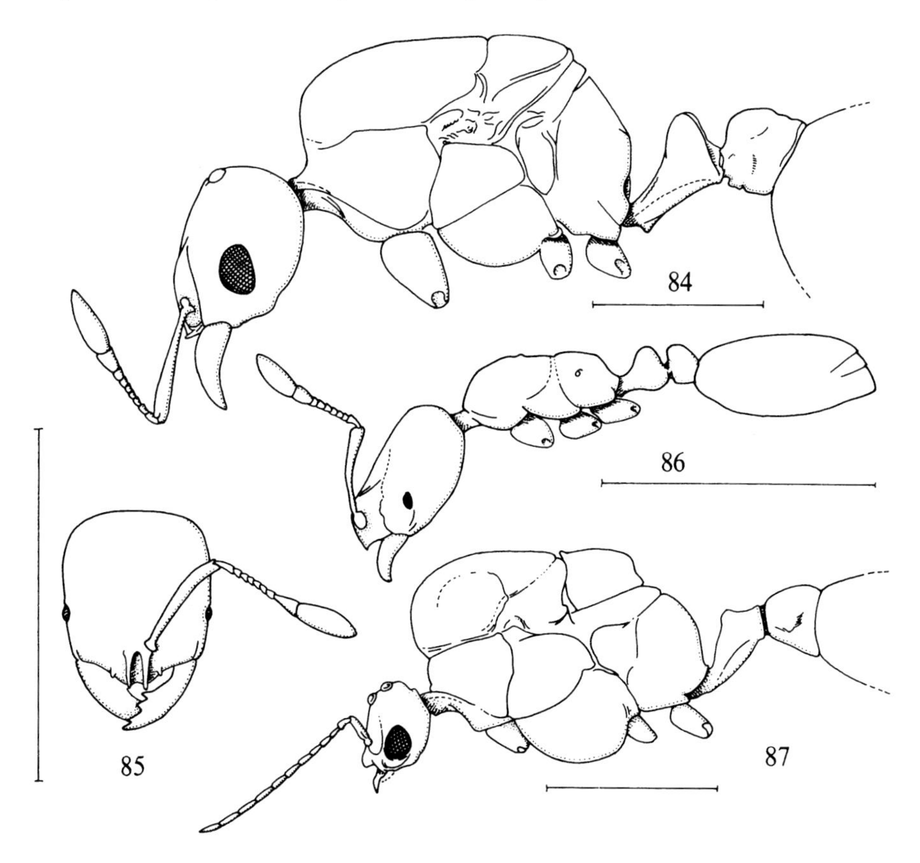

Figs. 84–87. *Diplorhoptrum fugax* (Latr.). - 84: queen in profile; 85: head of worker in dorsal view; 86: worker in profile; 87: male in profile. Scale: 1 mm.

TRIBE CREMATOGASTRINI FOREL

Genus *Crematogaster* Lund, 1831

Crematogaster Lund, 1831:132.
Type-species: *Formica scutellaris* Olivier, 1791.

All castes have the postpetiole attached to the dorsum of the first gaster segment. In the worker the cordate gaster is frequently carried uptilted and most species discharge a defensive deterrent fluid from the apical orifice, the sting being weak and atrophied. The male has very short antennal scapes, not longer than the two following funiculus segments.

This is a genus with many hundreds of species spread over the tropics and subtropics with a few palaearctic species none of which are endemic in North Europe.

Crematogaster scutellaris (Olivier, 1791)
Figs. 88, 89.

Formica scutellaris Olivier, 1791:497.

Worker. Head shining yellowish red contrasting with the brown alitrunk and dark gaster. Antennae 11 segmented. Length: 3.5–5.0 mm.

Queen. As worker, much larger. Length: 8.0–9.5 mm.

Male. Dark brown; antennae 12 segmented with very short scape. Mandibles reduced with three teeth. Length: 4.0–5.0 mm.

Biology. This arboricolous species is frequently introduced with cork from South Europe or North Africa and has on occasion established itself temporarily in and around warehouses and cork factories in England.

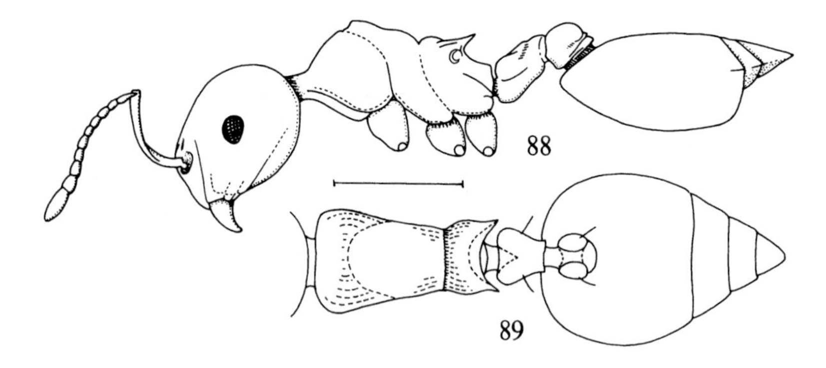

Figs. 88, 89. *Crematogaster scutellaris* (Oliv.). – 88: worker in profile; 89: alitrunk and gaster of worker in dorsal view. Scale: 1 mm.

TRIBE MYRMECININI ASHMEAD

[emend Emery, 1913]

Genus *Myrmecina* Curtis, 1829

Myrmecina Curtis, 1829:265.
Type-species: *Formica graminicola* Latreille, 1802.

Worker. Head square; clypeus bicarinate, the carinae terminating as two blunt teeth projecting from the anterior margin. Mandibles broad, fully denticulate, but leaving a free space basally when closed. Palp formula 4:3; antennae 12 segmented. The head has paired carinae ventrally. Propodeum with 2 small tubercules dorsally, anterior to the strongly developed propodeal spines. The holarctic members of this genus have the petiole and postpetiole quadrangular, joined broadly to propodeum and gaster respectively, without noticeable peduncle.

Queen. As worker but with more developed alitrunk; wings blackish, covered with short adpressed dark hairs.

Male. Antennal scape very short; petiole and postpetiole quadrangular; wings dark; mandibles very reduced, non-functional. Notauli distinct.

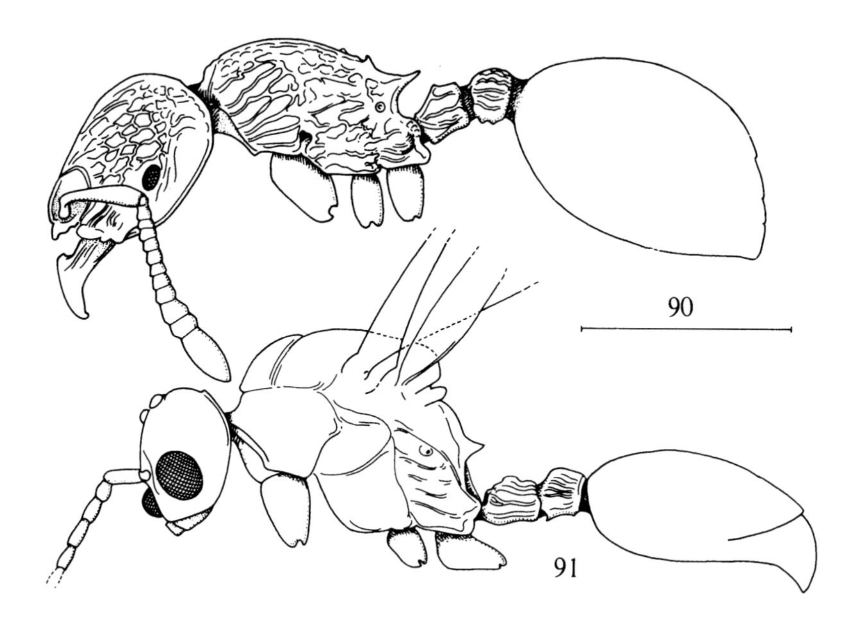

Figs. 90, 91. *Myrmecina graminicola* (Latr.). – 90: worker in profile; 91: male in profile. Scale: 1 mm.

The genus includes several Indo-Malayan and holarctic species of which one only occurs in North Europe.

16. ***Myrmecina graminicola*** (Latreille, 1802)
Figs. 90,91.

Formica graminicola Latreille, 1802:256.

Worker. Blackish brown with front of head, underside and appendages rusty yellow. Body and appendages strongly haired. Antennae with intermediate funicular segments transverse. Pronotum with angled antero-lateral corners. Head and alitrunk strongly rugose. Length: 3–3.6 mm.

Queen. As worker, often with more reddish areas exposed. Length: 4–4.2 mm.

Male. Black, smooth and shining, hairy. Eyes and ocelli large. Mandibles edentate, very reduced. Wings very dark, pilose; forewings with 1 cubital and 1 discoidal cell. Femora dilated in middle. Length: 3.4–4 mm.

Distribution. Very local in Denmark: EJ, and South Sweden: Sk., Sm., Öl., Gtl., Vg. – In British Isles rather local from South coast of England to Midlands and South Wales. – Range: Portugal to Caucasus, North Africa to Sweden.

Biology. This is a sluggish slow moving species; workers are often found individually in nests of other ant species and on disturbance tend to curl into a tight ball. Colonies occur under stones in stony pastures and in open woodland and may consist of several hundred workers with several queens and often including intermediate forms between worker and queen. This species is mainly scavenging and does not attend aphids. Alatae are developed during late summer and have been taken outside nests from August to October.

TRIBE LEPTOTHORACINI EMERY

Genus *Leptothorax* Mayr, 1855

Leptothorax Mayr, 1855:431.

Type-species: *Formica acervorum* Fabricius, 1793.

North European species small, worker length 2.3–4 mm. Body hairs clavate, not pointed; pronotum rounded anterolaterally; propodeal spines distinct; antennae 11 or 12 segmented in female castes, w ith 3 segmented club as long as rest of funiculus. Male has 12 or 13 antennal segments. Mandibles distinctly 5-toothed. Notauli very distinct.

In Europe, there are at least 40 species distinguishable on rather slight characters and their taxonomy is in need of revision. Seven species occur in Scandinavia. The North European species live in small communities of 30 up to 300 individuals under stones, in rock crevices, under bark, in twigs or in peat.

Keys to species of *Leptothorax*

Workers

1 Robust species with 11 segmented antennae and very distinct mesopropodeal furrow 2
– More slender species with 12 segmented antennae 3
2 (1) Tibiae and scapes with numerous suberect hairs, antennal club dark brown to nearly black, size larger, 3.2–4 mm (Fig. 92) .... 17. *acervorum* (Fabricius)
– Tibiae and scapes with occasional adherent hairs only, antennal club yellow brown to brown, size smaller, 2.5–3.5 mm (Fig. 93) ...... 18. *muscorum* (Nylander)
3 (1) Antennal clubs pale, concolorous with rest of antenna, distinct mesopropodeal furrow, propodeal spines long and stout (Fig. 96) .... 19. *nylanderi* (Förster)
– Antennal clubs brown to black, no mesopropodeal furrow 4
4 (3) Propodeal spines reduced to very short broad denticles, petiole outline a blunt triangle in profile, antennal clubs pale brown .... 20. *corticalis* (Schenck)
– Propodeal spines distinct, antennal clubs brown to black, distinctly darker than rest of funiculus 5
5 (4) Propodeal spines long and curved, petiole in profile sharply angled with anterior face rising steeply to anterodorsal ridge; gaster usually with dark band more or less interrupted in middle and front corners of head blackish (Fig.98) 21. *interruptus* (Schenck)
– Propodeal spines short; petiole with a short truncate dorsal area; dark patches on head and gaster not interrupted medially 6
6 (5) Dorsal outline of alitrunk somewhat flattened, often with a slight depression between pronotum and mesonotum; gaster usually with clearly defined dark band across base of first segment. Antennal clubs pale brown to dark brown; anterior and dorsal faces of petiole meet at a distinct angle (Fig. 100) 23. *unifasciatus* (Latreille)
– Dorsal outline of alitrunk convex without a break; gaster pale or dark but not banded; antennal clubs dark brown to black; dorsum of petiole rounded into anterior face without distinct angle (Fig. 99) .... 22. *tuberum* (Fabricius)

Queens

1 Antennae 11 segmented 2
– Antennae 12 segmented 3
2 (1) Scapes and tibiae with numerous suberect hairs .... 17. *acervorum* (Fabricius)
– Scapes and tibiae with sparse short adherent hairs .... 18. *muscorum* (Nylander)
3 (1) Antennal clubs pale brown, concolorous with rest of funiculus 4
– Antennal clubs distinctly darker than rest of funiculus 5
4 (3) Propodeal spines robust; petiole high with distinct dorsal area; gaster usually banded; alitrunk yellowish .... 19. *nylanderi* (Förster)

– Spines reduced to very short denticles; petiole low rising to an oblique angled peak without dorsal area; body colour including gaster evenly reddish brown
20. *corticalis* (Schenck)

5 (3) Mesoscutellum striate throughout; petiole with short truncate dorsal area; gaster not banded .... 22. *tuberum* (Fabricius)

– Mesoscutellum diffusely sculptured and shining; petiole angled or peaked in profile; gaster often distinctly banded 6

6 (5) Propodeal spines reduced to short denticles, shorter than half the space between; alitrunk yellowish brown .... 23. *unifasciatus* (Latreille)

– Propodeal spines well developed, as long as space between; alitrunk brown to dark brown .... 21. *interruptus* (Schenck)

Males

1 Antennae 12 segmented; scapes shorter than second funiculus segment........... 2

– Antennae 13 segmented; scapes longer than second funiculus segment 3

2 (1) Tibiae with numerous long suberect hairs; large robust species. Length 4–4.5 mm .... 17. *acervorum* (Fabricius)

– Tibiae bare or with very short hairs only; slender species. Length 3–3.5 mm (Fig. 95) .... 18. *muscorum* (Nylander)

3 (1) Space between notauli smooth; funiculus segments 2 to 5 twice as long as broad 4

– Space between notauli sculptured; funiculus segments 2 to 5 less than twice as long as broad 5

4 (3) Whole alitrunk smooth and shining .... 20. *corticalis* (Schenck)

– Sides of promesonotum finely striated .... 19. *nylanderi* (Förster)

5 (3) Antennal segments 2 to 5 not longer than broad; propodeal spines very distinct; body uniformly dark and closely sculptured .... 21. *interruptus* (Schenck)

– Antennal segments 2 to 5 slightly longer than broad; propodeal spines absent or reduced to faint tubercules; body colour pale to dark brown, general appearance more shining 6

6 (5) Area between notauli with rugulose sculpture throughout; petiole scarcely longer than high, ratio 10:8.2; scape as long as 4 following funiculus segments; appendages very pale brown (Fig. 101) .... 22. *tuberum* (Fabricius)

– Area between notauli with dilute sculpture, the striae almost obsolete in centre; petiole longer than high, ratio 10:7.2; scape as long as 3 following segments; appendages almost colourless .... 23. *unifasciatus* (Latreille)

17. ***Leptothorax acervorum*** (Fabricius, 1793)
Fig. 92.

Formica acervorum Fabricius, 1793:358.

Worker. Reddish to brownish yellow with the head, antennal club and dorsal surface of

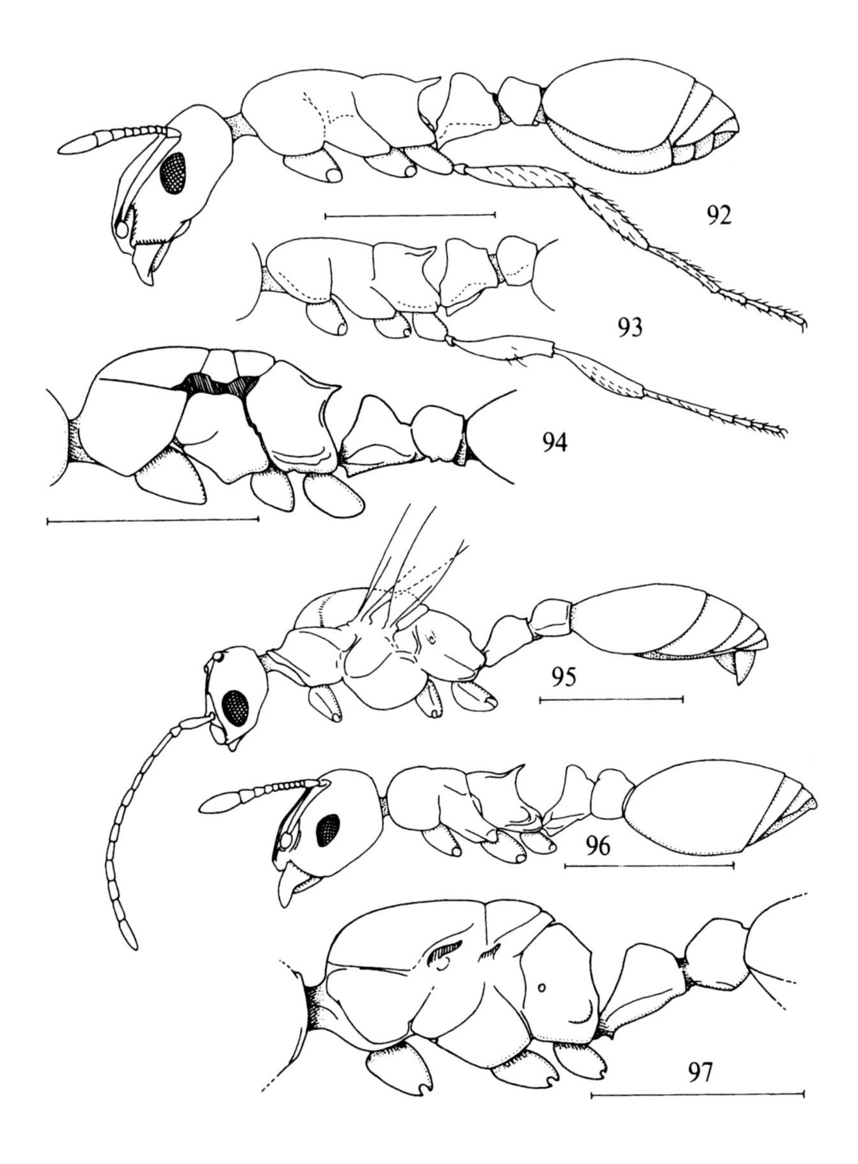

Figs. 92–97. *Leptothorax* spp. in profile. – 92: worker of *acervorum* (Fabr.); 93: worker of *muscorum* (Nyl.); 94: queen of same; 95: male of same; 96: worker of *nylanderi* (Förster); 97: queen of *corticalis* (Schenck). Scale: 1 mm.

gaster darker. Dorsa of petiole nodes and femora frequently infuscated. Antennae with eleven segments. Head longitudinally striate, alitrunk rugose and gaster smooth. Propodeal spines strong. Mesopropodeal suture distinct and depressed. Tibiae and scapes with numerous erect hairs. Length: 3.8–4.5 mm.

Queen. As worker but darker sometimes almost black. Length: 3.8–4.8 mm.

Male. Brownish black, large and robust; antennae 12 segmented with very short scape; semi-erect hairs numerous on tibiae. Length: 4.5–5.0 mm.

Distribution. Abundant throughout Denmark, Fennoscandia and British Isles. – Range: northernmost Scandinavia to mountains of South Europe and from Spain to Japan.

Biology. This species nests in small isolated colonies of 25 to 60 individuals with one or several queens; worker-queen intercastes are frequent. It is found nesting in open moorland in peat, rock crevices and under stones and in woodland areas on fallen tree trunks, rotten branches, stumps or under bark. The workers forage singly, predating small insects or scavenging insect corpses. It has not been observed to tend aphids, is non-aggressive and avoids combat with other ants. Alatae occur in the nests in June and July and have been observed flying and mating on high ground in July.

Note. This is a comparatively large and robust species easily recognised by the abundant suberect appendage hairs in all castes. The species tends to darken in colour from south to north varying from bright yellowish brown to nearly black, the darker samples occurring chiefly in high mountain areas, peat bogs and in the arctic north but with no clear break in colour gradation to the dark form sometimes referred to as the variety *nigrescens* Ruzsky (1905).

18. ***Leptothorax muscorum*** (Nylander, 1846)
Figs. 93–95.

Myrmica muscorum Nylander, 1846b:1054.

Worker. Pale brown to brown with antennal clubs and head often slightly darker. Propodeal spines short but distinct. General appearance more slender than *L. acervorum*. Head striate, alitrunk rugose and gaster smooth. Appendage hairs few and adpressed. Clypeus in some Scandinavian series have a distinct median concavity as described for *L. gredleri* Mayr (Buschinger, 1966) but they are not otherwise different from the typical species. Length: 2.4–3.2 mm.

Queen. As worker, normally darker. Length: 2.7–3.2 mm.

Male. Brownish black; occasional semi-erect hairs on tibiae but much shorter and less profuse than in *L. acervorum*. Antennae 12 segmented with very short scape.

Distribution. South to Central Fennoscandia not uncommon, till about latitude 63°. Local in Denmark. Absent from British Isles. – Range: Appenines to Central Scandinavia and Pyrenees to Urals, not found in British Isles but common in continental Europe.

Biology. This species is similar to *L. acervorum* but smaller and more slender. In Scandinavia, it is restricted to sheltered valleys in woodland areas where it nests in stony banks, tree stumps or under bark. Its habits are similar to those of *L. acervorum* with small colonies having one or occasionally two queens. Males and alate queens are found in July and August.

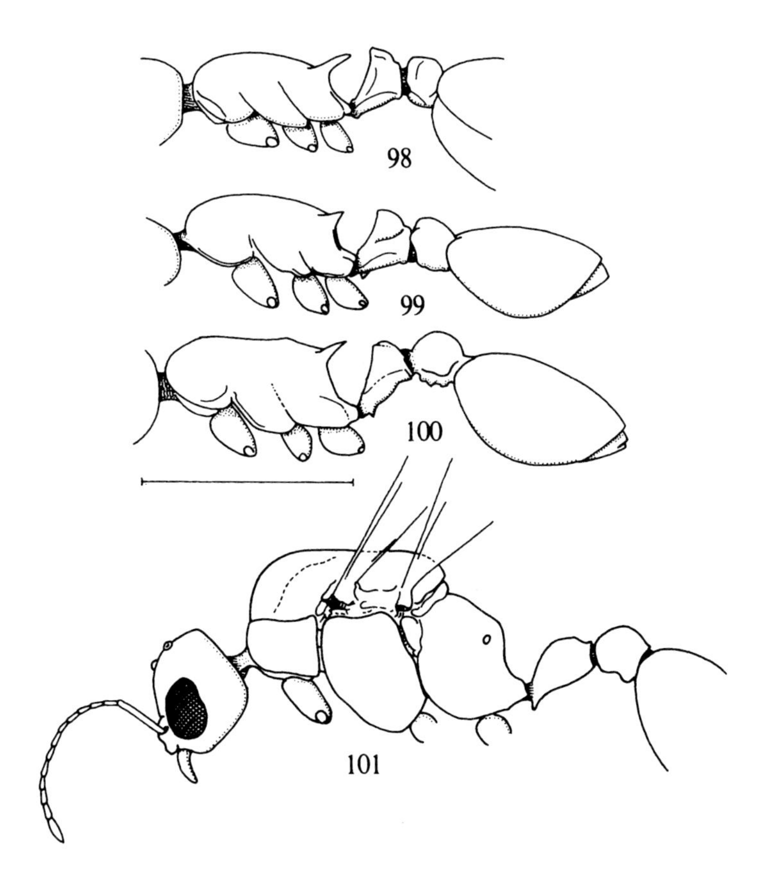

Figs. 98–101. *Leptothorax* spp. in profile. – 98: worker of *interruptus* (Schenck); 99: worker of *tuberum* (Fabr.); 100: worker of *unifasciatus* (Latr.); 101: male of *tuberum* (Fabr.). Scale: 1 mm.

19. ***Leptothorax nylanderi*** (Förster, 1850)
Fig. 96.

Myrmica nylanderi Förster, 1850:53.

Worker. Yellow to pale yellowish brown, the head sometimes darker, broadly infuscate on the first gaster segment. Antennae including clubs and legs concolorous with the rest of the body. Head longitudinally striate, alitrunk finely rugose, gaster smooth. Legs without erect hairs. Antennae twelve segmented; mesopropodeal impression distinct and clearly visible in side view. Length: 2.3–3.4 mm.

Queen. As worker, but with enlarged alitrunk and banded gaster. Length: 4.2–4.7 mm.

Male. Brownish black with pale yellowish appendages, mesonotum between the notauli, and most of the rest of the alitrunk, smooth and shining with some striae at the sides of the promesonotum. Antennae 13 segmented; funiculus segments 2–5 nearly twice as long as broad. Length: 3.0–3.2 mm.

Distribution. Sweden: Gotska Sandön and Gotland, and on the mainland from a few of the southern districts. – Not found in the other Scandinavian countries. – Not uncommon in parks and woodland in South England. – Range: Central and South Europe from Spain to Caucasus, north to South Sweden.

Biology. This species is immediately distinguishable from all other North European *Leptothorax* with twelve segmented antennae in the worker caste by the distinct mesopropodeal suture seen as a clear depression in the dorsal profile. It is normally a woodland bark inhabiting species but occasionally is found under stones. Its morphology and biology in France where it is common, has been intensively studied by Plateaux (1970). The species is normally monogynous with between 100 and 200 workers. Is is somewhat more aggressive than *L. acervorum* and despite its small size will attack and sting freely. Alate queens and males are developed during July and flights occur during early August.

20. ***Leptothorax corticalis*** (Schenck, 1852)
Fig. 97.

Myrmica corticalis Schenck, 1852:100.

Worker. Pale reddish brown with the head and gaster usually darker, antennae unicolorous brown; alitrunk and head finely longitudinally striate but general appearance shining. Propodeal spines reduced to very short denticles; petiole node triangular in profile with dorsal area reduced or absent. Length: 2.5–3.2 mm.

Queen. As worker, length 3.5–4.0 mm.

Male. Alitrunk smooth with no striae at sides; antennae pale brown; segments 2–3 only very slightly longer than broad. Length: 3.0 mm.

Distribution: Very local in Sweden: Vstm., Upl. (Forsslund, 1957a); Öl. (Douwes,

1976a). - Not found in Britain. - Range: Central Europe, rare, recorded only from Spain, France, North Italy, Austria, Czechoslovakia, Germany and Belgium.

Biology. This is a little known rare Central European species. It is a bark inhabitant and Forsslund (1957a) found it nesting in old oak trees inhabited by *Lasius brunneus* (Latr.). Only worker samples were seen. Douwes (1976a) records workers taken on Öland on oak and in a fallen branch. The species appears to be closely related to *L. nylanderi* which has similar habits.

21. ***Leptothorax interruptus*** (Schenck, 1852)
Fig. 98.

Myrmica interrupta Schenck, 1852:106.

Worker. Light bright yellow with dark areas at the side of the dorsum of the first gaster segment and frequently at the front corners of the head. The antennal club is distinctly dark. The dorsal outline of the alitrunk is smoothly curved without a break and the propodeal spines are long and curved. The petiole node is steeply peaked in profile. Length: 2.3–3.4 mm.

Queen. Uniformly dark with gaster often banded, middle of scutellum unsculptured, smooth. Length: 3.7–4.2 mm.

Male. Dark with very pale appendages and shortened funiculus segments - nos 2 to 5 are only very slightly longer than wide. Tibiae and scapes have no erect hairs. Length: 2.5–3.0 mm.

Distribution. Sweden: Gtl. and G. Sand. only. There is an old unverified record for Østfold in Norway. - In Britain locally in Kent, Wight, Hants and Dorset. - Range: sparsely distributed from Spain to Czechoslovakia and North Italy to Sweden.

Biology. This is a rather uncommon ground nesting species. In Britain where it has been well studied by Donisthorpe (1927) it is found nesting in dry peat or among small stones and heather roots in small colonies of 50–100 workers and single queens. Alatae are present in the nests during July.

22. ***Leptothorax tuberum*** (Fabricius, 1775)
Figs. 99, 101.

Formica tuberum Fabricius, 1775:393.

Workers. Colour varies from entirely pale yellowish brown with the head pale to almost black and the dorsum of the gaster brownish. The antennal clubs are brown to brownish black contrasting with the rest of the funiculus. The head and clypeus are longitudinally striate and the alitrunk rugose. The petiole node has a distinct but short truncate dorsal area; propodeal spines are very short but quite distinct. Length: 2.3–3.4 mm.

Queen. Brown to brownish black with scutellum striate throughout, rest as worker. Length: 3.7–4.5 mm.

Male. Brownish black; appendages very pale; antennal scape as long as 4 following segments. Space between notauli rugulose. Length: 2.5–3.2 mm.

Distribution. Common in South Norway, Sweden and Finland north to about latitude 62°, local in Denmark and in the coastal counties of S. England. – Range: a common and very widely distributed species in the mountains of Central Europe from Spain to the Caucasus and North Italy to Central Sweden.

Biology. This species characteristically nests in small single queened colonies under stones and in rock crevices. In Scandinavia it is restricted to warm lowland habitats. The alatae are found in July and August.

Note. The original and very brief description was based on Swedish material but the types are lost. Most Scandinavian samples have dark heads and could be referred to the supposed species *L. nigriceps* Mayr, 1855. In England where *L. tuberum* is locally abundant along the south coast, the colour tends to be uniformly pale but samples also occur with dark heads and there are no structural differences between dark headed and light coloured series.

23. ***Leptothorax unifasciatus*** (Latreille, 1798)

Fig. 100.

Formica unifasciatus Latreille, 1798:47.

Worker. Yellowish with the head often brown and the gaster characteristically having a dark band across the first gaster segment. The propodeal spines are short but strong and distinct. The outline of the alitrunk is more flattened than in *L. interruptus* and the petiole node in the worker has a distinct dorsal truncate area which meets the anterior face at a clearly defined angle. Length: 2.8–3.5 mm.

Queen. Pale brown to brownish black with distinct dark band across first gaster tergite; mesoscutellum diffusely sculptured; antennal clubs brownish, darker than rest of funiculus. Propodeal spines reduced to very short denticles; petiole more peaked than in worker. Length: 4.0–4.5 mm.

Male. Brownish black, appendages very pale; antennal scape as long as following 3 segments; space between notauli with dilute sculpture. Length: 2.8–3.5 mm.

Distribution. Very local, recorded from Sweden: Öl. only, first taken in 1968 by H. Andersson and later rediscovered among limestone ruins by P. Douwes (1976a). – The species does not occur in England but is common on the Channel Islands. – Range: an abundant species throughout Central and South Europe from Spain to the Caucasus and from South Italy to North Germany and the Netherlands.

Biology. This widely distributed and common species is rather similar to *L. tuberum* but is slightly larger and in the female castes easily distinguished by the banded gaster. It nests among rocks and under stones but also frequently occurs under bark. Individual colonies are usually more populous than the other North European

Leptothorax species with workers numbering 200 or more. Alatae are found in July and August.

Genus *Formicoxenus* Mayr, 1855

Formicoxenus Mayr, 1855:413.
Type-species: *Myrmica nitidula* Nylander, 1846.

Worker and queen. Small, shining, narrow bodied ants. Antennae 11 segmented; palp formula 4:3. Postpetiole with peg-like ventral process directed forward.

Males. Apterous, ergatoid without wing sclerites. Antennae 12 segmented. Postpetiole as in female castes.

This is a palaearctic genus with 2 species only; a similar species *Leptothorax diversipilosus* M. Smith, occurs in North America.

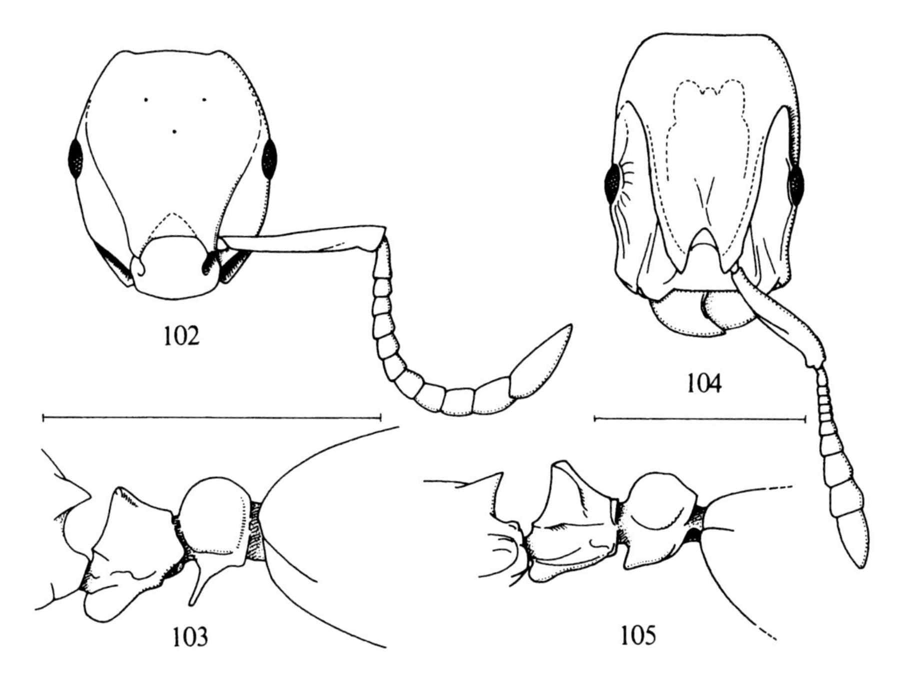

Figs. 102, 103. *Formicoxenus nitidulus* (Nyl.). – 102: head of worker in dorsal view; 103: petiole and postpetiole of worker. Scale: 1 mm.

Figs. 104, 105. *Harpagoxenus sublaevis* (Nyl.). – 104: head of worker in dorsal view; 105: petiole and postpetiole of worker. Scale: 1 mm.

24. ***Formicoxenus nitidulus*** (Nylander, 1846)
Figs. 102, 103.

Myrmica nitidula Nylander, 1946b:1056.

Worker. Reddish yellow to brown; whole surface of body smooth and shining with scattered acute pale hairs. Antennal club 3 segmented as long as rest of funiculus; propodeal spines short, set horizontally. Length: 2.8–3.4 mm.

Queen. Very like the worker, somewhat darker; eyes larger and ocelli present. Forewings with 1 cubital cell and 1 discoidal cell with open radial cell. Length: 3.4–3.6 mm.

Male. Worker like but with additional antennal and gaster segments; ocelli present. Antennae with funiculus relatively longer, terminating in 4 segmented club. Length: 2.8–3.2 mm.

Distribution. Throughout Denmark and Fennoscandia; local in England and Scotland. – Range: Spain to Eastern Siberia, North Italy to latitude 70° N.

Biology. This ant occurs only in the nests of *Formica rufa* and allied mound building species. It is ignored by its host among which the inquiline species moves freely. Individual nests contain only a few individuals, up to about 100, but often several nests are present within one mound of the host. Nests are located in fragments of wood, hollow twigs, bases of old bracken stems and in the earth floor of the *Formica* mound. Intermediate forms between queen and worker are common and individuals normally remain concealed within the nests but may wander on the mound surface on warm dull days. It is not known to feed on the *Formica* brood but in captivity will destroy *Leptothorax* larvae. Males and winged females may be found during July and August, mating occurring on the surface of the *Formica* mound.

Genus *Harpagoxenus* Forel, 1893

Harpagoxenus Forel, 1893: 167.
Type-species: *Myrmica sublaevis* Nylander, 1849.

Workers and queens. Head large, subrectangular with broad edentate mandibles. Both petiole and postpetiole have strongly developed ventral tooth-like projections. Palp formula 5:3. Antennae 11 segmented.

Male. Similar to *Leptothorax acervorum* but with ventral projecting tooth on postpetiole as in female castes. Antennae 12 segmented.

This genus contains at most 2 or 3 palaearctic and 1 North American species living in semi-parasitic dulotic relationship with members of the closely related genus *Leptothorax.*

25. ***Harpagoxenus sublaevis*** (Nylander, 1849)
Figs. 104, 105.

Myrmica sublaevis Nylander, 1849:33.

Worker. Pale yellowish brown to brown; head large, rectangular, with weakly concave occiput. Frontal carinae extend backward to enclose whole length of antennal scape. Antennae 11 segmented with intermediate segments strongly transverse and enlarged 4 segmented club. Eyes large, set midway at sides of head. Mesopropodeal furrow deep and distinct; propodeal spines broad and short. Femora and tibiae short and broadly rounded. Head and mesopropodeum longitudinally striate, petiole nodes and gaster smooth and shining. Whole body and appendages covered with long, acute, pale hairs. Length: 3.5–5.5 mm.

Queen. Ergatoid, similar to worker, but larger. Length: 4.7–5.7 mm. Normal alate queen has the head more square and the alitrunk relatively more massive, straightsided from above. Wings pale yellowish, short, forewings with open radial cell, 1 cubital cell and 1 discoidal cell. Length: 4.5–4.8 mm.

Male. Dark brown to black with paler legs and gaster. Head broader than long with rounded occiput. Mandibles edentate, short, reduced and nonfunctional. Eyes large, set anterior to midline of head, approximately $^2/_5$ length of head. Ocelli present but inconspicuous. Antennal scape short, less than 2 following segments. Head and alitrunk finely sculptured but whole body moderately shining. Length: 3.7–4.5 mm.

Distribution. Denmark: only recorded from Jutland; throughout Fennoscandia, not uncommon. – Absent from British Isles. – Range: Pyrenees to Caucasus; North Italy to North Norway.

Biology. This species lives in obligate dulotic association with *Leptothorax acervorum, L. muscorum* and more rarely with *L. tuberum.* Workers may forage singly outside the nest and are capable of brood tending and can feed themselves but are evidently dependant on the host species for the continuation of the colony. In Denmark and Fennoscandia nests containing host species and inquiline are commonly found in twigs on the ground, tree stumps or under bark but in the mountains of Central Europe they occur more usually under stones. In mixed colonies the host queen may survive and alatae of both species may be developed within the same nest. Fertilised *Harpagoxenus* queens invade new nests of *Leptothorax* and appropriate mature larvae and pupae of the host species to use both as food and for rearing as auxiliaries.

TRIBE TETRAMORIINI WHEELER

[emend Bolton, 1976]

Genus *Anergates* Forel, 1874

Anergates Forel, 1874:32

Type-species: *Myrmica atratula* Schenck, 1852.

This worker-less parasite genus contains one species only found locally throughout Eurosiberia and in some eastern states of North America.

26. ***Anergates atratulus*** (Schenck, 1852)
Figs. 106, 107.

Myrmica atratula Schenck, 1852:91.

Queen. Blackish brown with yellow appendages and mandibles. Mandibles reduced with single apical tooth. Palp formula 1:1, palps reduced. Clypeus with anterior margin broadly incised. Antennae 10 or 11 segmented. Eyes large, set median laterally; ocelli present. Fore-wings with 1 discoidal cell and open radial cell. Petiole transverse, postpetiole twice as broad as long, broadly attached to gaster. Gaster with longitudinal furrow in virgin queens, strongly physogastric in mature queens. Length: 2.5 mm.

Male. Pupoidal and apterous, dull pale grey. Mandibles lobiform, edentate. Antennae 10 or 11 segmented; ocelli present. Clypeus with anterior margin incised. Alitrunk with flight sclerites but wings never developed. Petiole and postpetiole compressed and broadly attached to following segments. Apex of gaster reflexed ventrally. Genitalia large and prominent. Length: 2.3 mm.

Distribution. Rare in Denmark and South Sweden, recorded only from NEZ, B, Hall., Öl. and Gtl. – In England recorded locally from Devon, Dorset, Hants and Surrey. – Range: Spain to Central Siberia, North Italy to South Sweden, also North America.

Biology. This is an obligate parasite of *Tetramorium caespitum*. Queens fertilised within the nest of the host species fly away to secure adoption in other colonies. Queens of the host species are not present in *Anergates-Tetramorium* nests and only *Anergates* brood are developed, often in large numbers, from the single adopted *Anergates* queen whose gaster becomes grossly swollen.

Genus *Strongylognathus* Mayr, 1853

Strongylognathus Mayr, 1853:389.
Type-species: *Eciton testaceum* Schenck, 1852.

Worker and queen. Mandibles falcate, tapering to apex, edentate or occasionally with a minute denticle before apex. Palp formula 4:3, antennae 12 segmented. Propodeum with a pair of very small denticles.

Male. Mandibles as worker but smaller; antennae 10 segmented with elongated second funicular segment as in *Tetramorium*.

This is a palaearctic genus with several species each dependant on one or other *Tetramorium* species. One species only occurs in North Europe.

27. ***Strongylognathus testaceus*** (Schenck, 1852)
Figs. 108, 109.

Eciton testaceum Schenck, 1852:117.

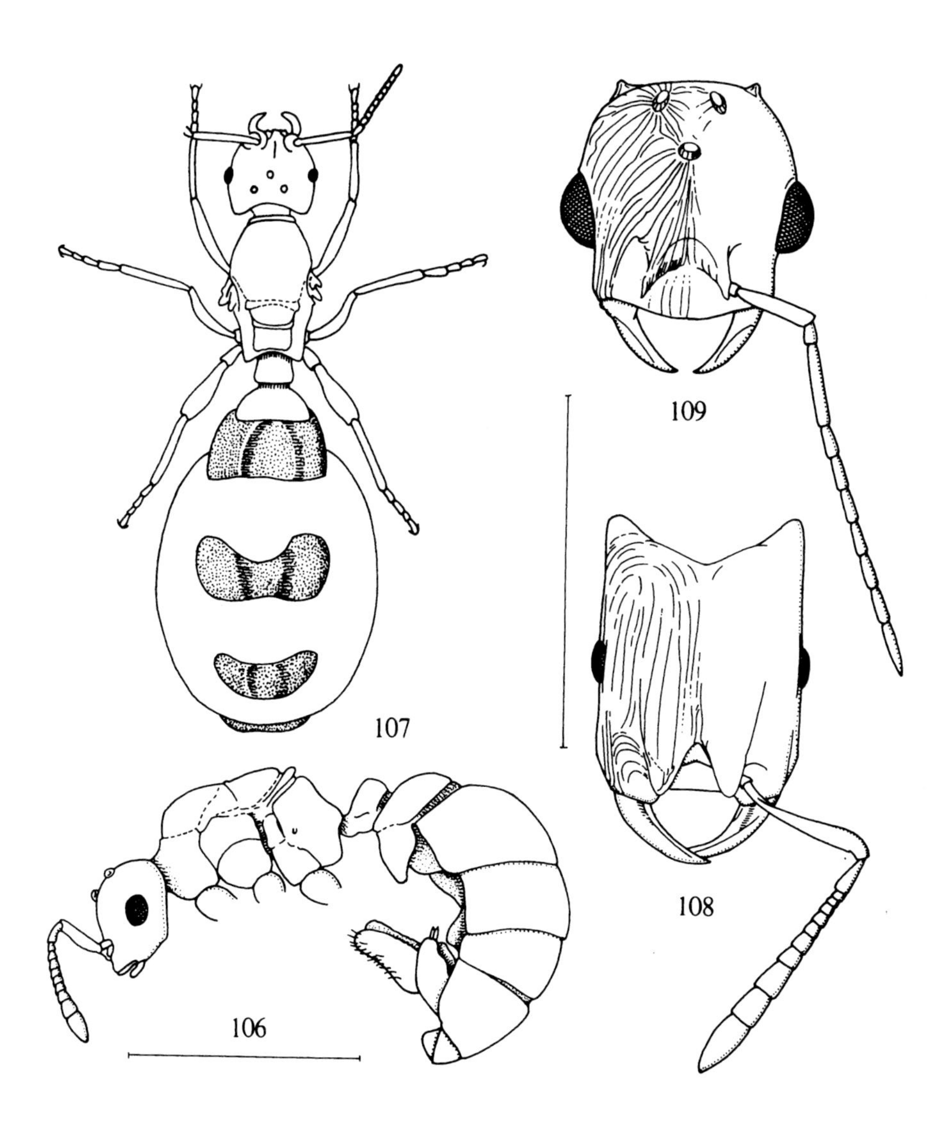

Figs. 106, 107. *Anergates atratulus* (Schenck). – 106: male in profile; 107: queen in dorsal view. Scale: 1 mm.
Figs. 108, 109. *Strongylognathus testaceus* (Schenck). – 108: head of worker in dorsal view; 109: head of male in dorsal view. Scale: 1 mm.

Worker. Yellowish brown. Head rectangular with pronounced occipital emargination and postero-lateral angles. Body shining with long fine pale hairs present also on appendages. Sculpture variable, with longitudinal striae present or more or less effaced on head and alitrunk. Length: 2.0–3.6 mm.

Queen. Darker than worker, brownish with paler appendages. Wings pale with 1 cubital cell, 1 discoidal and an open radial cell. Length: 3.5–3.8 mm.

Male. Dark brown with pale brown appendages. Head small narrower than promesonotum. Occiput emarginate sharply angled postero-laterally. Antennal scape shorter than second funiculus segment. Length: 3.2–4 mm.

Distribution. Very rare in S. Sweden, only recorded from Sm. and Öl. – Recorded in England from Devon, Dorset and Hants very locally. – Range: Pyrenees to Ukraine, North Italy to Sweden.

Biology. This species occurs only in the nests of its host *Tetramorium caespitum.* Workers and brood of both host and parasite are present but only the sexuals of *Strongylognathus* are developed, the original *Tetramorium* queen as well as the adoptive *Strongylognathus* queen usually being found present together.

S. testaceus workers are normally greatly outnumbered by *Tetramorium* workers. Observations on this and related species suggest that neighbouring nests of the host species are raided to recruit more *Tetramorium* pupae to the colony which is often very populous with up to 20,000 individuals. Alatae are present in July and August.

Genus *Tetramorium* Mayr, 1855

Tetramorium Mayr, 1855:423.

Type-species: *Formica caespitum* Linné, 1758.

Worker and queen. Lateral margins of clypeus raised into a ridge in front of antennal insertions; mandibles with 3 strong apical teeth followed by a row of smaller denticles. Sting with lamelliform appendage. Pronotum of worker distinctly angled anterolaterally. Body hairs simple.

Male. Antennae 10 segmented with elongate second funiculus segment, mandibles dentate. Body hairs simple.

This is a very large genus with several hundred species mainly distributed throughout the palaeotropic and palaearctic regions. One species is native in North Europe and one or more subtropical cosmopolitan species are occasionally introduced and may become established in heated premises.

Keys to species of *Tetramorium*

Workers and queens

1 Frontal carinae short; body colour brownish black .... 28. *caespitum* (Linné)

- Frontal carinae extended backwards as longitudinal ridges almost to occipital margin; body colour yellowish to reddish brown 2

2 (1) Dorsum of alitrunk and petiole nodes coarsely rugulose; body hairs long and numerous. Queen has first gaster tergite with fine longitudinal striae at base (Fig. 113) .... *bicarinatum* (Nylander)

- Alitrunk finely rugulose with numerous punctures; body hairs short and sparse. Queen has postpetiole and first gaster tergite finely punctulate *simillimum* (Smith)

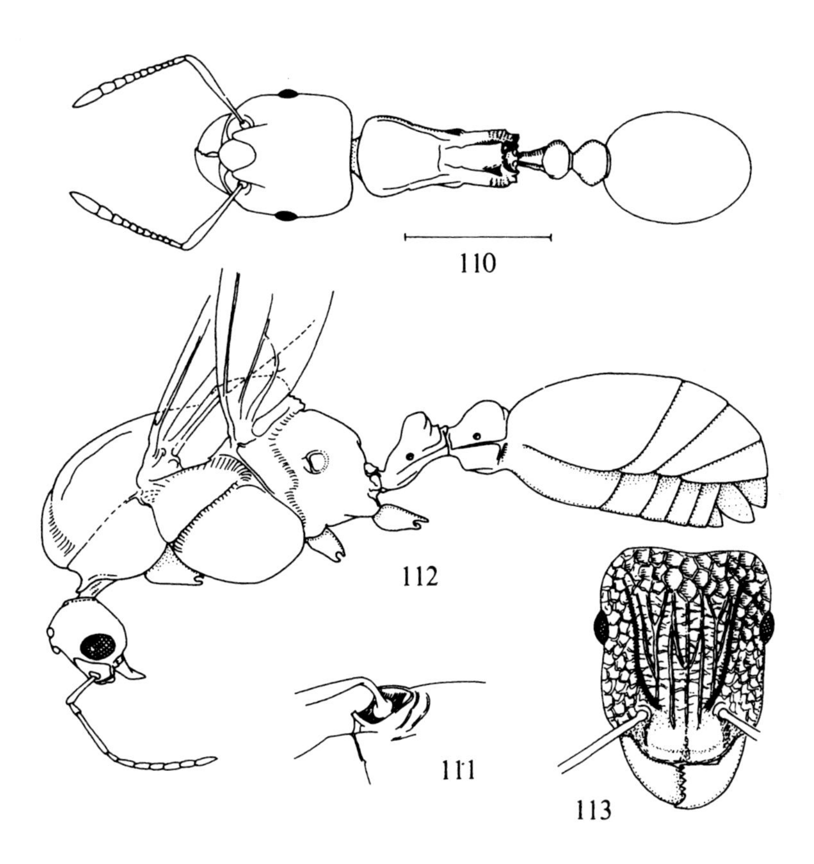

Figs. 110–112. *Tetramorium caespitum* (L.). - 110: worker in dorsal view; 111: antennal insertion in anterolateral view, showing raised clypeal border; 112: male in profile. Scale: 1 mm.

Fig. 113. *Tetramorium bicarinatum* (Nyl.), head of worker in dorsal view.

Males

1 Postpetiole smooth and shining .... *simillimum* (Smith)
– Postpetiole punctured or regulose 2
2 (1) Head and alitrunk yellowish brown. Frontal carinae extend backward to level of ocelli .... *bicarinatum* (Nylander)
– Head and alitrunk blackish brown. Frontal carinae short 28. *caespitum* (Linné)

28. ***Tetramorium caespitum*** (Linné, 1758)
Figs. 2, 110–112.
Formica caespitum Linné, 1758:581.

Worker. Blackish brown, sometimes paler; head including clypeus and alitrunk regularly longitudinally striate. Petiole and postpetiole with shallow punctures and sculpture but smooth in centre. Propodeal spines very short, broadly denticulate, petiole and postpetiole about as broad as long. Length: 2.5–4 mm.

Queen. Blackish brown with appendages and mandibles paler. Pronotum concealed above by overarching mesonotum. Mesonotum and scutellum smooth and shining. Much larger than worker with petiole and postpetiole broadly transverse. Wings pale with 1 discoidal and 1 cubital cell and open radial cell; pterostigma and veins yellowish. Length: 6–8.0 mm.

Male. Head much narrower than alitrunk, rounded with very large eyes; antennal scape shorter than second funiculus segment. Y-shaped notauli and parapsidal furrows distinct. Postpetiole much wider than long. Head, propodeum, petiole and postpetiole longitudinally striate, mid body more finely striate. Size much larger than worker. Length: 5.5–7 mm.

Distribution. Locally common in Denmark and Southern Fennoscandia up to approximately latitude 62° 50'. – Range: holarctic: America to Japan, North Africa to N. Europe including British Isles.

Biology. The species tends to be coastal in North Europe but also inland on heath and on the open borders of woodland, nesting in the earth and also under stones. Colonies are normally single queened, but populous with up to 10,000 or more workers. This species is moderately aggressive, living by predation on other arthropods, scavenging and also from root aphid honeydew. Seeds of various herbs and grasses are often collected into the nest. The alatae are conspicuously large compared with the workers; they are developed in the early summer and fly in late June and July.

Tetramorium bicarinatum (Nylander, 1846) [stat. rev. Bolton, 1977]
Fig. 113.

Myrmica bicarinata Nylander, 1846b:1061.

Worker and queen. Pale reddish, coarse reticulo-rugulose sculpture; frontal carinae prolonged backward; propodeal spines long. Length of worker: 3.4–4 mm, queen: 5–5.5 mm.

Male. Yellow brown to brown; mesonotum and postpetiole shining, rest of alitrunk and head weakly sculptured; frontal carinae prolonged backward; occiput bluntly angled at posterolateral borders; propodeum with 2 short spines. Length: 4.5–5 mm.

Biology. This is a cosmopolitan species of tropical origin often introduced and established in heated glasshouses in the British Isles. It nests in small communities in earth, under bark and in or on shrubby hothouse plants. Long know as *Tetramorium guineense* (Fabricius), Bolton (1977) has shown that the correct name is *T. bicarinatum*.

Tetramorium simillimum (Smith, 1851)

Myrmica simillima Smith, 1851:118.

Workers and queen. Weakly sculptured, pale red much smaller than *T. guineense*. Length of worker: 1.6–2 mm, queen: 2.2–2.5 mm.

Male. Light yellowish red; occiput rounded. Length: 2.5 mm.

Biology. This cosmopolitan species occasionally occurs in heated glasshouses in Europe and has been recorded from Denmark and also on several occasions in England.

SUBFAMILY FORMICINAE LEPELETIER

Workers and queens with ventral apex of gaster (hypopygium) produced into a conical structure terminating in a circular acidopore fringed with hairs. Petiole a large scale or distinct node. Gaster with 5 distinct tergites visible in dorsal view. Males have semi-erect hairs on dorsum of alitrunk.

Keys to genera of Formicinae

Workers and queens

1 Antennal insertions set at a distance behind posterior clypeal margin; metapleural gland orifice absent (Figs. 114, 116) .... *Camponotus* Mayr (p. 86)
– Antennal insertions more or less contiguous with posterior clypeal margin. Metapleural gland orifice present 2
2 (1) Antennae 11 segmented (Fig. 158) .... *Plagiolepis* Mayr (p. 110)
– Antennae 12 segmented 3
3 (2) Eyes at or in front of midlength of sides of head; petiole inclined forward, overhung by first gaster tergite (Fig. 156) .... *Paratrechina* Motschulsky (p. 108)
– Eyes behind midlength of sides of head; petiole nodal or as a vertical scale not overhung by gaster 4
4 (3) Mandibles falcate, pointed (Fig. 267) .... *Polyergus* Latreille (p. 155)
– Mandibles with broad masticatory border, coarsely dentate 5

5 (4) Propodeal spiracle ellipsoid or slitlike set at a distance from posterior propodeal declivity. Funiculus segments 2–5 as long or longer than segments 6–10 (Figs. 159, 176) .... *Formica* Linné (p. 111)
– Propodeal spiracle circular or broadly oval set close to posterior margin of propodeum. Funiculus segments 2–5 shorter than segments 6–10 (Figs. 124, 135) *Lasius* Fabricius (p. 92)

Males

1 Antennal insertions set at a distance behind posterior clypeal margin (Fig. 119) *Camponotus* Mayr (p. 86)
– Antennal insertions set close to or at posterior clypeal margin 3
2 (1) Eyes set in front of or at midlength of sides of head. Gonopalpi absent *Paratrechina* Motschulsky (p. 108)
– Eyes set behind of sides of head. Gonopalpi present 3
3 (2) Antennae 12 segmented .... *Plagiolepis* Mayr (p. 110)
– Antennae 13 segmented 4
4 (3) Mandibles very reduced, falcate. Antennal scapes shorter than first four following segments. Maxillary palps 4 segmented very reduced *Polyergus* Latreille (p. 155)
– Mandibles broadening to apex. Antennal scapes longer than first five following segments. Maxillary palps 5–6 segmented 5
5 (4) Propodeal spiracle narrowly elliptical, set well forward from the posterior propodeal margin (Fig. 192) .... *Formica* Linné (p. 111)
– Propodeal spiracle broadly oval or circular, set close to or at the posterior propodeal margin (Fig. 129) .... *Lasius* Fabricius (p. 92)

TRIBE CAMPONOTINI FOREL

Genus *Camponotus* Mayr, 1861

Camponotus Mayr, 1861:35.
Type-species: *Formica ligniperda* Latreille, 1802.

This is a world wide genus with a large number of species reaching their greatest abundance in the tropics. The form of the alitrunk and head varies considerably. Although attempts have been made to differentiate species groups, it has not been possible to make clearcut distinctions in all cases to justify the use of subgeneric names. Despite the variety of form, the attachment of the antennal scape some distance from the clypeal border is a constant feature that immediately distinguishes the genus from *Formica* and *Lasius*. The antennae are 12 segmented in the female and worker, 13 in the male; segments 2 to 5 are marginally longer than those following. Maxillary palps 6 segmented, labial palps 4 segmented. Frontal carinae are sinuate broadening behind the antennal insertions. Ocelli are absent in the worker, small but distinct in the female and male. Wings with one discoidal cell, cubital cell absent. Male external genitalia small.

In North Europe the 4 Fennoscandian species all mine in dead wood but only *C. herculeanus* occasionally mines in live trees. This and *C. ligniperda* are among the largest ants found in Europe with major workers up to 12 or 14 mm long and females up to 18 mm long.

Keys to species of *Camponotus*

Workers

1 Front clypeal border incised in middle (Fig. 114) .... 29. *fallax* (Nylander)
– Front clypeal border entire 2
2 (1) Colour uniformly black; pubescence thick; projecting hairs profuse over whole body including gaster (Fig. 118) .... 30. *vagus* (Scopoli)
– Colour in part reddish; pubescence thin; hairs on gaster sparse mainly restricted to tergite borders 3
3 (2) Gaster shining with pubescence short, sparse often absent over medial areas of first and second gaster tergite. Basal face of first tergite and sometimes whole tergite reddish; alitrunk bright yellowish red to dark red
32. *ligniperda* (Latreille)
– Gaster somewhat dull with long pubescence evenly distributed over dorsal surface; basis of first gaster tergite often with a small reddish patch behind scale; alitrunk dull red, sometimes reddish black .... 31. *herculeanus* (Linné)

Queens

1 Front clypeal border incised in middle; size smaller: head width 2.0–2.2mm
29. *fallax* (Nylander)
– Front clypeal border entire; size larger: head width over 3 mm 2
2 (1) Colour uniformly black with thick pubescence and numerous standing hairs on gaster .... 30. *vagus* (Scopoli)
– Bicoloured with at least propodeum reddish 3
3 (2) General appearance shining, pubescence short sparse or absent over medial area of first gaster tergite (Fig. 122). Punctuation on frons shallow
32. *ligniperda* (Latreille)
– General appearance somewhat dull, gaster pubescence long and evenly distributed over surface (Fig. 123). Punctuation on frons deep
31. *herculeanus* (Linné)

Males

1 Front border of clypeus with shallow emargination .... 29. *fallax* (Nylander)
– Front border of clypeus convex 2

2 (1) Long hairs abundant over dorsum of head and gaster; dorsal crest of petiole sharply angled at sides enclosing wide and deep emargination (Fig. 115) ... 30. *vagus* (Scopoli)

– Head and dorsal surface of gaster with sparse hairs; petiole with shallow emargination and sides of dorsal crest more rounded (Fig. 121) ... 3

3 (2) From above, gaster fringed at the sides with scattered projecting hairs; pubescence long (0.075–0.125 mm) ... 31. *herculeanus* (Linné)

– From above, first two gaster tergites usually without hairs; pubescence sparse and short (0.05 mm) ... 32. *ligniperda* (Latreille)

29. ***Camponotus fallax*** (Nylander, 1856)
Figs. 114, 115.

Formica fallax Nylander, 1856:57.

Worker. Dark brownish red to black with legs and antennae paler; body hairs sparse; microsculpture on head and alitrunk dense, giving somewhat opaque appearance; gaster shining. Clypeus not projecting forward beyond mandibular insertions, middle of front border incised; in the larger examples the cleft is deep, giving a bidentate appearance. Mandibles broad with five distinct teeth. In profile dorsum of alitrunk rather flat, propodeum with steep descending basal face; petiole broadly oval in front view. Length: variable 4–9 mm.

Queen. Similar in appearance with long steeply descending basal face of propodeum. Length: 8–10 mm.

Male. Brownish black; clypeal emargination shallow, sometimes indistinct; petiole shallowly emarginate, low and thick in profile. Wings pale except for yellowish front border and stigma of fore-wing. Mandibles with apical tooth only. Length: 7–8 mm.

Distribution. Sweden: Västmanland, where Forsslund (1957) found it locally in old oak trees. Very rare. – Range: Central and South Europe, Portugal to Ukraine and Morocco to Poland.

Biology. This species lives in small colonies of 30–50 individuals under bark or in dead wood of old trees up to 2 m or more above ground in open deciduous woodland

---

Figs. 114, 115. *Camponotus fallax* (Nyl.). – 114: head of worker in dorsal view; 115: petiole scale of male in posterior view. Scale: 1 mm.

Figs. 116–120. *Camponotus vagus* (Scop.). – 116: head of major worker in dorsal view; 117: head of minor worker in dorsal view; 118: worker in profile; 119: male in profile; 120: petiole scale of male in posterior view. Scale: 1 mm.

Fig. 121. *Camponotus herculeanus* (L.), petiole scale of male.

Figs. 122, 123. Gaster tergite 1 in queens of *Camponotus*. – 122: *ligniperda* (Latr.); 123: *herculeanus* (L.). Scale: 1 mm.

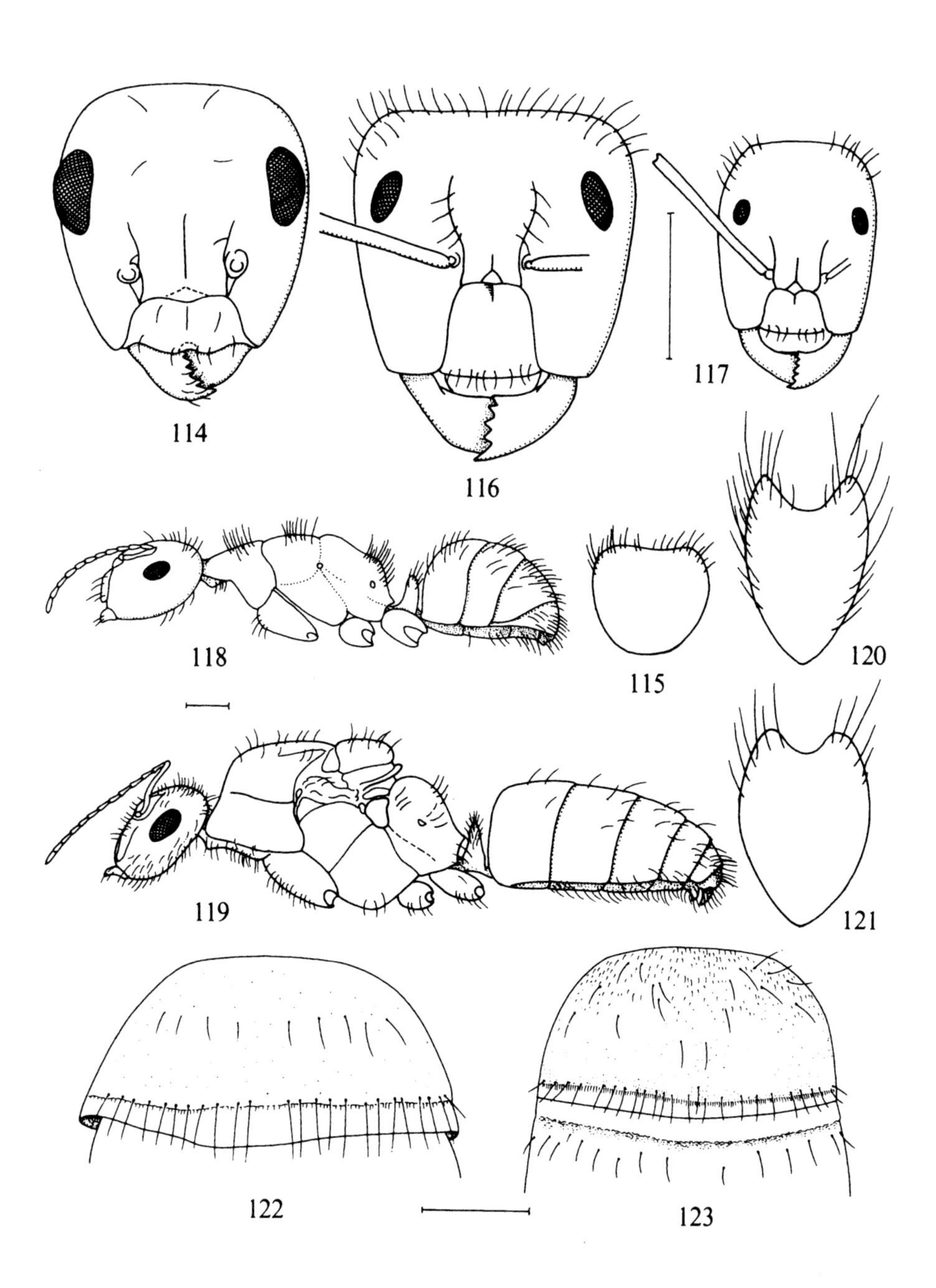
114
116
117
118
115
120
119
121
122
123

or parkland. Workers forage singly and are fugitive. Alatae have been recorded in early summer, May and June.

30. ***Camponotus vagus*** (Scopoli, 1763)
Figs. 116–120.

Formica vaga Scopoli, 1763:312.

In this group of species the anterior border of the clypeus is entire, straight or feebly convex and does not extend beyond the mandibular insertions. The alitrunk in the worker caste is high and steep sided; in profile the dorsum is convex without a break, the dorsal face of the propodeum abruptly curving into the long almost vertical basal face. From above the pronotum is much wider than the rest of the alitrunk which narrows to half its width posteriorly. Mandibles are large with five strong teeth which are often blunted and worn in the larger workers. The male has the mandibles slender with an apical tooth only.

Worker. Uniformly black with profuse body hairs. The sculpture is finely transverse and closely punctured, obscured by long thick pubescence. Length: 6–12 mm.

Queen. As worker. Length: 14–16 mm.

Male. Pubescence dilute; petiole deeply emarginate rising to a sharp acute angle at each side of the dorsal crest. Length: 9–10 mm.

Distribution. Sweden: Öl. and Gtl. – Finland: Ab and Ka. Rare. – Range: a South European species abundant in the Mediterranean area, but recorded from Portugal to South Russia and the mountains of North Africa to Poland.

Biology. *C. vagus* nests in dry rotten wood among roots under stones in dry sun exposed banks. It is an active aggressive species biting freely on disturbance. As with other species of this group it is both carnivorous and aphidicolous. According to Pisarski (1961) alatae have been recorded in July in Poland where it occurs very locally in the Centre and South.

Note. Forsslund (1957a) doubted its existence in Fennoscandia as the only verifiable specimen was an alate queen taken in Öland. However, there is a good series of workers from Gotland in the University of Lund collection, an old specimen from Karelia australis in the Helsinki Museum and a live colony has been kept under observation by A. K. Merisuo at Rymättyllä in southwest Finland (Merisuo and Käpylä, 1975).

31. ***Camponotus herculeanus*** (Linné, 1758)
Figs. 121, 123.

Formica herculeana Linné, 1758:579.

Workers. Bicoloured with alitrunk dull red to reddish black, head and gaster dull black.

Frons has deep close set punctures. Long pubescence on dorsum of first gaster tergite overlapping posterior border. Length: 5–12 mm.

Queen. Sides of alitrunk, propodeum and scale dull red, rest black; punctures and pubescence as in worker. Length: 14–17 mm.

Male. Entirely black, tarsi and funiculi paler; wings yellowish infuscate; head and gaster hairs sparse but present on all gaster tergites. Petiole scale has shallow dorsal emargination with raised sides rounded to blunt lateral angles.

Distribution. Throughout Fennoscandia; local in Denmark and absent from the British Isles. – Range: throughout mountain Europe and extending through Northern Eurasia from Norway to Eastern Siberia to the northernmost tree frontier in Arctic Norway.

Biology. This species is a typical denizen of shaded coniferous forest nesting in rotten stumps and occasionally mining in living trees. Fertilised females found nests singly. Alatae are developed in the late summer but overwinter to swarm in June.

32. ***Camponotus ligniperda*** (Latreille, 1802)
Fig. 122.

Formica ligniperda Latreille, 1802a:88.

This species is similar in all castes to *C. herculeanus* but distinguished by the brighter colour and more shining gaster.

Worker. Alitrunk bright yellowish red to red; pubescence is short and sparse, usually absent on medial dorsal surfaces of the first and second gaster tergites. Length: 6–14 mm.

Queen. Propodeum, scale and usually anterior face of gaster clear red, rest as worker, head punctures shallow. Length: 16–18 mm.

Male. Pilosity reduced so that projecting hairs usually absent from the first gaster tergite with only one or two present towards the posterior border of the second tergite. Length: 8–12 mm.

Distribution. Common in Central and South Fennoscandia north to latitude 63°; more local in Denmark; not found in British Isles. – Range: Central Spain to West Russia, Sicily to Central Sweden.

Biology. This species is characteristically found in stony banks and along the sun exposed borders of woodland, either nesting under stones or in dry stumps. It is an aggressive ant biting freely and will sometimes attack other *Camponotus* or *Formica* colonies. The larger workers bite their opponents clean through the alitrunk or crush their heads with their strong mandibles. A more xerothermic species than *C. herculeanus* its habits are otherwise similar.

TRIBE LASIINI EMERY

Genus *Lasius* Fabricius, 1804

Lasius Fabricius, 1804: 415.
Type-species: *Formica nigra* Linné, 1758.

The antennae are 12 segmented in the worker and female, 13 segmented in the male. The antennal insertions are situated at the posterior margin of the clypeus and segments 2 to 5 are not longer than the succeeding segments. Maxillary palps are 6 segmented, labial palps 4 segmented. The head of the worker caste is somewhat cordate with a posterior emargination in some species. The clypeus is broad and rounded anteriorly. The frontal carinae are short and sub-parallel and in most species the frontal triangle is indistinctly defined. The orifice of the propodeal spiracle is circular or broadly oval situated close to the posterior propodeal border. Ocelli are minute or indistinct in the worker but distinct in the female and male. The petiole is vertical and scale-like in most species.

This genus contains about 42 species with a holarctic distribution. It was extensively revised by Wilson (1955). Members of this genus are predominantly aphidicolous but also carnivorous and scavenging. There are 14 species in Europe of which 10 are known to occur in Fennoscandia.

Keys to species of *Lasius*

Workers

1 Colour shining black; head large, broadly cordate with a distinct posterior emargination (Fig. 137) .... 37. *fuliginosus* (Latreille)
– Colour greyish or brownish black or yellow 2
2 (1) Colour yellow to brownish yellow; maximum eye length $^1/_6$ head width or less ... 3
– Colour brownish or greyish black or somewhat bicoloured reddish and black; maximum eye length more than $^1/_5$ head width 8
3 (2) Petiole nodal with dorsal crest in front view strongly convex; head emarginate posteriorly; with genal margins rounding in towards close set mandibular insertions (Figs. 153, 155) .... 42. *carniolicus* (Mayr)
– Petiole with thin dorsal crest, straight or emarginate in front view; back of head convex and genal margins not pronouncedly sloping in towards mandibular insertions which are wide set 4
4 (3) Front tibiae and antennal scapes with sub-erect hairs standing out clearly from general pubescence 5
– Front tibiae and antennal scapes with pubescence only 6
5 (4) Scapes and tibiae flattened with thin front edge; petiole scale narrowly rectangular .... 39. *meridionalis* (Bondroit)
– Scapes and tibiae rounded elliptical in cross section; petiole scale with curving sides and more or less emarginate dorsal border .... 38. *umbratus* (Nylander)

6 (4) Body hairs short, erect hairs on first gaster tergite × 0.3 or less hind tibial width (Fig. 149) .... 41. *mixtus* (Nylander)
- Body hairs long, erect hairs on first gaster tergite × 0.7 or more hind tibial width 7

7 (6) Petiole narrow with deep semicircular emargination; erect hairs on gaster restricted to hind borders of tergites (Fig. 147) .... 40. *bicornis* (Förster)
- Petiole broad with dorsal crest widely emarginate or straight; erect hairs arise all over dorsum of gaster (Fig. 124) .... 33. *flavus* (Fabricius)

8 (2) Front tibiae and antennal scapes with abundant semi-erect hairs
36. *niger* (Linné)
- Front tibiae and scapes bare (or with occasional oblique hairs standing out from general pubescence) 9

9 (8) Body bicoloured with head and alitrunk pale brownish red contrasting with darker gaster; occipital corners without projecting hairs; ocelli usually visible and frontal furrow clearly demarcated (Fig. 131) .... 35. *brunneus* (Latreille)
- Body evenly brownish or greyish black; occipital corners with projecting hairs; ocelli not visible and frontal furrow usually indistinct (Fig. 130)
34. *alienus* (Förster)

Queens

1 Colour shining black; orifice of metapleural gland without guard hairs; head broadly emarginate and wider than alitrunk; scutum overhangs the pronotal convexity (Fig. 136) .... 37. *fuliginosus* (Latreille)
- Colour various from yellowish brown to brownish or greyish black; orifice of metapleural gland with guard hairs; pronotal convexity not covered by scutum 2

2 (1) Front tibiae and antennal scapes with standing hairs 3
- Front tibiae and scapes with pubescence but no standing hairs 5

3 (2) Head distinctly narrower than alitrunk at its widest point; eyes without short hairs between facets; colour greyish black .... 36. *niger* (Linné)
- Head broader than alitrunk; eyes with short hairs; body colour yellowish brown to brownish black 4

4 (3) Antennal scapes and tibiae flattened with thin front edge: minimum hind tibial width × 0.5–0.6 maximum width. Petiole scale rectangular in front view; funiculus segments distinctly elongate; sculpture fine and pubescence thin so that general appearance, especially frons, shining; colour brownish black (Figs. 144, 145) .... 39. *meridionalis* (Bondroit)
- Antennal scapes and tibiae oval: minimum hind tibial width × 0.75 maximum width. Petiole scale with rounded sides; hexagonal in frontal view, with distinctly emarginate dorsal crest; sculpture and pubescence somewhat coarse so that general appearance somewhat dull; colour yellowish brown to dark mahogany brown (Figs. 140, 141) .... 38. *umbratus* (Nylander)

5 (2) Eyes without short hairs between facets, or one or two at most 6
- Eyes with numerous short hairs 7

6 (5) Body colour greyish black; head distinctly narrower than alitrunk; frontal triangle usually indistinct; wings clear, not infuscated .... 34. *alienus* (Förster)
– Body colour brownish black; head more massive, nearly as broad as maximum width of alitrunk; median furrow and frontal triangle always clearly demarcated; wings infuscated basally .... 35. *brunneus* (Latreille)
7 (5) Head distinctly narrower than alitrunk .... 33. *flavus* (Fabricius)
– Head as broad or broader than alitrunk 8
8 (7) Petiole very convex in front view; head with rounded occipital lobes and convex genal margins .... 42. *carniolicus* (Mayr)
– Petiole with thin scale-like dorsal crest; back of head straight or weakly concave; genal margins straight or very slightly convex 9
9 (8) Gaster with short hairs only, × 0.3 maximum tibial width or less; scale weakly emarginate; size larger – length: 6.0–7.5 mm .... 41. *mixtus* (Nylander)
– Gaster with long hairs, as long as tibial width; scale deeply incised; size smaller – length: 4.5–5.5 mm .... 40. *bicornis* (Förster)

Males

1 Suberect hairs present on either extensor tibial surface or antennal scapes or both 2
– Tibiae and scapes with pubescence only 4
2 (1) Mandibles with apical tooth only, masticatory border smoothly rounded into pre-apical cleft; head distinctly narrower than alitrunk .... 36. *niger* (Linné)
– Masticatory border with distinct teeth; head massive relative to alitrunk, as wide or wider 3
3 (2) Frons shining with fine microsculpture and thin pubescence; frontal groove and frontal triangle well marked; mandibles with very well defined teeth; body colour evenly black; cross vein m-cu frequently absent on fore-wing (Fig. 146) 39. *meridionalis* (Bondroit)
– Frons somewhat dull with coarse microsculpture and thick pubescence; frontal groove and triangle often indistinctly defined or obscured by pubescence; denticles less sharply defined; body colour brown to brownish black; cross vein m-cu usually present (Fig. 142) .... 38. *umbratus* (Nylander)
4 (1) Mandibles with a single apical tooth 5
– Mandibles with a distinct pre-apical tooth or denticles as well as an apical tooth 7
5 (4) Shining black; head large; distinctly emarginate posteriorly; metapleural gland lacking guard hairs (Fig. 138) .... 37. *fuliginosus* (Latreille)
– Colour grey to brownish black; head border convex or straight; metapleural gland with guard hairs 6
6 (5) Projecting hairs absent on occipital corners of head; pre-apical cleft of mandible clear, wings fuscous on basal half (Fig. 132) .... 35. *brunneus* (Latreille)
– Head above eyes fringed with projecting hairs; pre-apical cleft of mandibles shallow; basal angle of mandible broadly rounding into edentate masticatory border; wings clear .... 34. *alienus* (Förster)

7 (4) Petiole thickened in side view, with broadly rounded dorsal crest; back of head with numerous projecting hairs (Fig. 154) .... 42. *carniolicus* (Mayr)

– Petiole thin in side view, with emarginate or flat dorsal crest; back of head with occasional hairs only 8

8 (7) Head width less than maximum alitrunk width; mandibles with apical and one pre-apical tooth only; cross vein m-cu often absent on one or both fore wings (Fig. 127) .... 33. *flavus* (Fabricius)

– Head width as wide as alitrunk; mandibles either denticulate evenly or with at least one or more denticles in addition to apical and pre-apical teeth; cross vein m-cu normally present 9

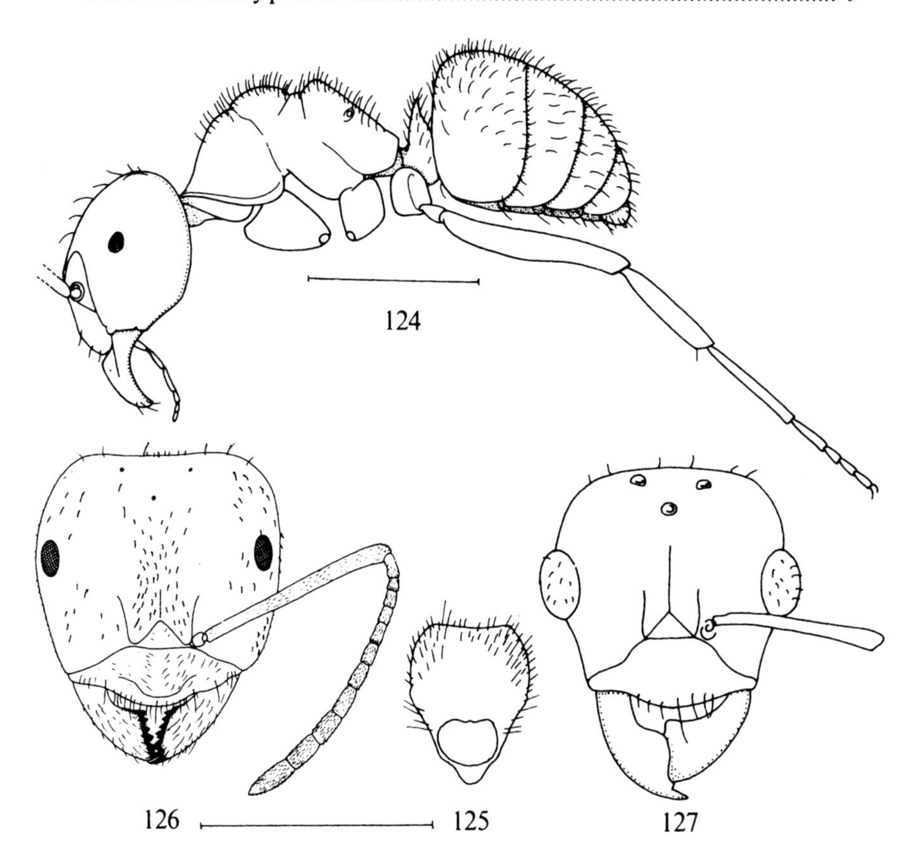

Figs. 124–127. *Lasius flavus* (Fabr.). – 124: worker in profile; 125: petiole scale of queen in anterior view; 126: head of worker in dorsal view; 127: head of male in dorsal view. Scale: 1 mm.

9 (8) Petiole scale high, narrow, and deeply incised; masticatory border with well defined denticles; size small – length: 3.2 mm .... 40. *bicornis* (Förster)

– Petiole scale broad, with straight or slightly emarginate dorsal crest; masticatory border with one or two denticles only, the rest obscure or absent; larger – length: 3.7–4.5 mm (Fig. 152) .... 41. *mixtus* (Nylander)

33. ***Lasius flavus*** (Fabricius, 1781)
Figs. 124–127.

Formica flava Fabricius, 1781:491.

Worker. Clear yellow to brownish yellow. Body hairs on dorsum of gaster and alitrunk long; appendages and body covered with more or less thick adpressed pubescence, more dilute on head. No erect hairs on tibiae, scapes or genae. Scale thin in side view, low and broad in front view with dorsal margin mildly convex straight or in larger specimens occasionally emarginate. Size very variable in North European populations. Length: 2.2–4.8 mm.

Queen. Light to dark brown with underside paler. Pubescence and pilosity as in worker. Head distinctly narrower than alitrunk. Eyes with numerous short hairs. Wings partly infuscate. Length: 7.2–9.5 mm.

Male: Dark brown to brownish black. Scape and tibial hairs entirely lacking; head narrower than alitrunk; mandibles with one apical and pre-apical tooth. Vein m-cu often missing on fore-wing but usually present in extreme northern populations. Wings faintly tinted but not infuscate. Length: 3.5–5.0 mm.

Distribution. Throughout Denmark and Southern Fennoscandia up to latitude 67°; one record for Polmark in the Norwegian Finnmark. – Throughout British Isles excluding Northern Islands. – Range: North America to Japan; North Africa to Arctic.

Biology. This species is very widely distributed and one of the most abundant in North Europe where it is a characteristic earth mound builder in pastures and along the periphery of woodlands but also nesting under stones in rocky areas. Colonies are started by one or more queens with primary pleometrose quite frequent. In North Europe nests in exposed places and in northern extremity of its distribution, *L. flavus* exhibits a wide range of worker size. On warm sites in southern areas usually in sandy lowland heath, worker size is small and much less variable. Eye ommatidium number is correlated with size and series of small workers with eyes with low ommatidium number are sometimes referred to *Lasius myops* Forel. However, queen size is constant regardless of worker size. *L. myops* is therefore regarded as a synonym of *L. flavus*. Individual nests may contain several thousand individuals and favourable nest sites, e. g. pasture sloping with a southern aspect, may be crowded with mound nests. This species, as with *L. niger*, tends to swarm on the same day in any one area and in years of abundant production of sexuals huge mating may occur during late July or August. This species is hypogoeic, seldom occurring above ground, feeding on small insects and the exudate of subterranean root feeding aphids.

34. ***Lasius alienus*** (Förster, 1850)
Figs. 128–130.

Formica aliena Förster, 1850:36.

Worker. Greyish yellow to brownish black. Pubescence adpressed, moderately thick over whole body and appendages. Short erect hairs scattered over dorsum and round whole occipital margin of head. Back of head convex. Ocelli indistinct or invisible;

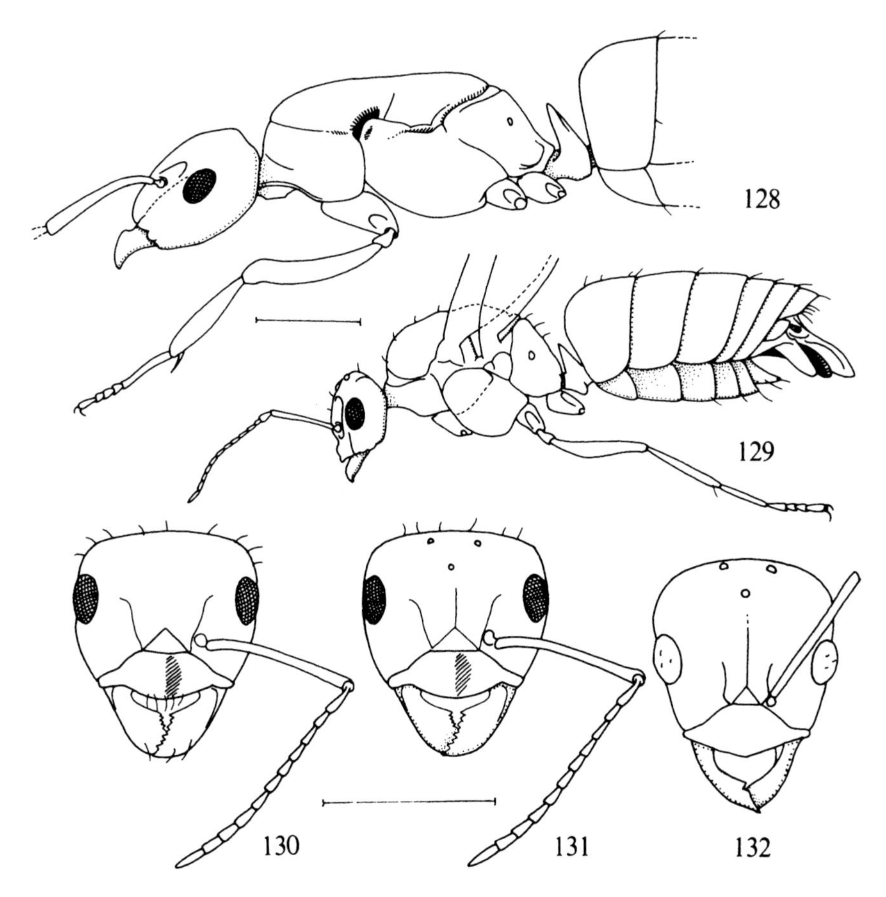

Figs. 128–130. *Lasius alienus* (Förster). – 128: head and alitrunk of queen in profile; 129: male in profile; 130: head of worker in dorsal view. Scale: 1 mm.

Figs. 131, 132. *Lasius brunneus* (Latr.). – 131: head of worker in dorsal view; 132: head of male in dorsal view. Scale: 1 mm.

frontal furrow indistinct. Erect hairs absent on scape and front tibiae, sometimes present on hind tibiae. Length: 3.0–4.2 mm.

Queen. Greyish black to brownish black. Wings hyaline. Head much narrower than broad alitrunk. Pubescence and pilosity as in worker, no erect hairs on scape and front tibiae. Length: 8.0–9.0 mm.

Male. Greyish black. Wings hyaline. No erect hairs on tibiae or scapes. Length: 3.0–3.8 mm.

Distribution. Common in Denmark and South Sweden; in Norway only recorded from Ø, HE and VE; in Finland recorded as far north as Ob and Ks. – Common in South Britain, local in Ireland and Southwest Scotland. – Range: Portugal to Japan also Himalayas, North Africa to Finland. Also North America (Wilson, 1955).

Biology. This wide ranging species nests in the soil on sandy lowland heaths, dry open pasture, sea cliffs and rocky outcrops in North Europe. Its habits are mainly subterranean, feeding on the exudates of root aphids but also by scavenging and predating small insects. Workers are generally unobtrusive and non aggressive compared with *Lasius niger*. Nests are single queened founded by solitary fertilised queens. Mating swarms occur in August.

35. ***Lasius brunneus*** (Latreille, 1798)
Figs. 131, 132.

Formica brunnea Latreille, 1798: 41.

Worker. Bicoloured with gaster dark brown contrasting with testaceous or pale reddish brown head and alitrunk. Pubescence and body hairs sparse. Occipital hairs restricted to median area of back of head only. Scapes and tibiae never with erect hairs. Back of head flat or feebly concave. Frontal triangle and frontal furrow distinct, ocelli small but always clearly visible. Length: 3.2–4.5 mm.

Queen. Dark brown. Head comparatively massive, nearly as broad as alitrunk. Fore-wings partly infuscate. Ocelli conspicuous. Pilosity and other head features as worker. Head width 1.50–1.65 mm. Length: 8.0–9.0 mm.

Male. Dark brown. Head large with conspicuous frontal suture; mandibles with well marked pre-apical cleft. Dorsal margin of petiole deeply concave. Fore-wings partly infuscate. Length: 3.5–4.5 mm.

Distribution. Denmark: LFM, local; Sweden: Sm. – Dlr; Norway: Ø, B, AK, VE and TE. – Not recorded from Finland. – Locally common in Central South England. – Range: Spain to Crimea and West Himalayas, Italy to Sweden.

Biology. This species nests in the interior of old trees, chiefly oak, but has also been recorded from hedgerows. It is fugitive and non-aggressive, rapidly dispersing on disturbance and because of its cryptic habits may be somewhat under-recorded. In Norway and Sweden it has frequently occurred nesting in the timbers of old houses and farm buildings, where its populous colonies may be difficult to dislodge. It chiefly tends tree aphids including the large bark feeding *Stomaphis*. Single queens initiate colonies in the crevices of old trees but may also be accepted back into the mother nest after the mating flight which occurs in June and early July.

36. ***Lasius niger*** (Linné, 1758)
Figs. 133–135.

Formica nigra Linné, 1758:580.

Worker. Greyish brown to dark brownish black, mid body occasionally somewhat paler. All appendage surfaces including scapes and tibiae with abundant erect hairs. Length: 3.5–5.0 mm.

Queen. Brownish black. Scape and tibial hairs abundant. Wings hyaline. Alitrunk massive relative to head which is always narrower. Length: 8.0–9.0 mm.

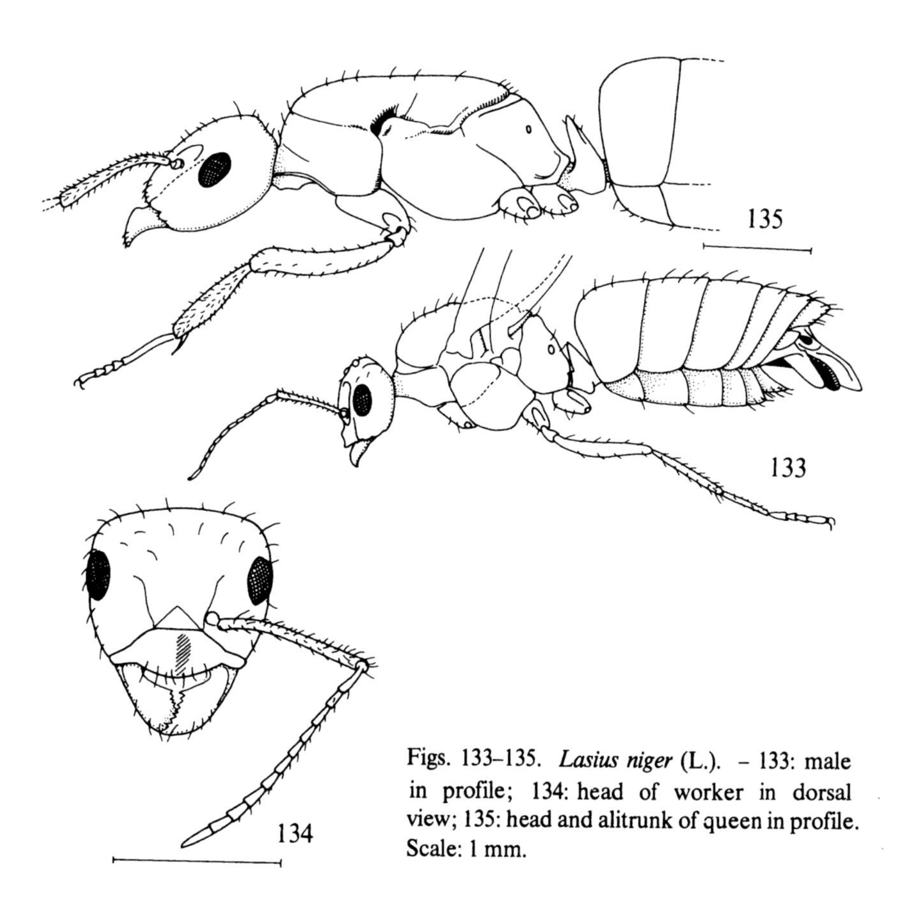

Figs. 133–135. *Lasius niger* (L.). - 133: male in profile; 134: head of worker in dorsal view; 135: head and alitrunk of queen in profile. Scale: 1 mm.

Male. Brownish black. Mandibles with single apical tooth with slight central depression on masticatory border. Erect hairs variable, usually less abundant than in queen, but always some present on tibiae. Wings hyaline. Length: 3.5–5.0 mm.

Distribution. Throughout Denmark, Fennoscandia and British Isles as far north as latitude 64°; abundant in all southern areas. – Range: Holarctic from Western United States to Japan, North Africa to Finland.

Biology. This is one of the commonest European species and is often especially evident at sites of human disturbance including towns, villages and quarries. Nests may occur in walls, pavements, tree stumps in open woodland, pasture and open heath. Occasionally earth mounts are formed and foraging tracks are frequently covered by surface tunnels of earth. This species is aggressive and readily attacks other ants. Nests are single queened and moderately populous with several hundred up to 10,000 workers. Aphids on shrubs and herbs as well as subterranean species are tended. Mating swarms occur from July to late August and in some years huge numbers may fly over a large district on the same date.

Lasius emarginatus (Olivier, 1791:494)

This is a Central and South European species that occurs in Poland, and the Channel Islands. It is distinguished in the worker caste by the distinctly red alitrunk, more sparse, oblique appendage hairs and relatively longer antennal scapes. The queen has the mesoscutum reddish and distinctly flattened. The male has the mesopleurae in part testaceous yellow, sparse scape hairs and a more sculptured frontal triangle than *L. niger*.

37. ***Lasius fuliginosus*** (Latreille, 1798)
Figs. 136–138.

Formica fuliginosa Latreille, 1798:36.

Worker. Shining black, legs brownish yellow; pubescence sparse, scattered erect hairs over dorsum. Head broadly cordate, emarginate posteriorly with rounded occipital lobes; genital margins incurving towards mandibular insertions. Maxillary palps short with segments 4, 5 and 6 subequal. Petiole thickened wedge shaped in profile, with feebly convex faces, dorsal margin narrow, convex or straight. Length: 4.0–6.0 mm.

Queen. Colour and shape as worker. Pubescence and body hairs thicker and more abundant than worker. Head width: 1.45–1.65 mm, broader than alitrunk. Length: 6.0–6.5 mm.

Male. Shining black; head cordate, not emarginate posteriorly, as wide as alitrunk. Petiole low and thick with rounded dorsal margin. Mandibles with apical tooth only. Length: 4.5–5.0 mm.

Distribution. Throughout Denmark and Southern Fennoscandia to latitude 62°; South Ireland, England and Wales. – Range: Portugal to Japan and North India, South Italy to Finland.

Biology. This distinctive species is easily recognised by its shining black colour and broad head. Carton nests are constructed at the base of old trees, hedgerows and sometimes in sand dunes and in old walls. Colonies are populous, often polycalic with more than one focal nest and several queens. Workers forage above ground in narrow files throughout the day and night during warm weather, ascending trees and shrubs to tend aphids. The mandibles are relatively weak but small insects may be taken as food. Other competing ant species are repelled by aromatic anal secretions. Fertilised queens may be retained in the old nest or found fresh colonies through adoption by the members of the *Lasius umbratus* species group; mixed colonies with *L. umbratus* or *L. mixtus* have often been observed. Flight periods are irregular and have been recorded in all months from May to October. A number of local beetles occur with this species including members of the genus *Zyras* which exhibit protective mimicry. Waldén (1964), records an enormous nest measuring 63 x 55 x 55 cm found in a cellar near Göteborg and there are similar reports from outbuildings and cellars in England (Donisthorpe, 1927).

38. ***Lasius umbratus*** (Nylander, 1846)
Figs. 139–142.

Formica umbrata Nylander, 1846b: 1048.

Worker. Clear yellow to reddish yellow; funiculus segments 2 to 4 slightly longer than

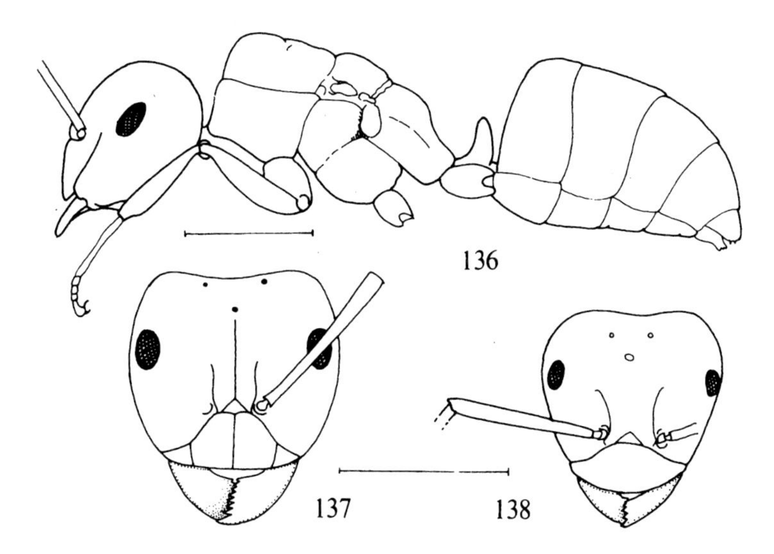

Figs. 136–138. *Lasius fuliginosus* (Latr.). – 136: queen in profile; 137: head of worker in dorsal view; 138: head of male in dorsal view. Scale: 1 mm.

wide. Scape elliptical in cross section. Petiole tapering to dorsal crest which is usually emarginate. Body surface and appendages covered in adpressed silvery pubescence. Longest hairs on gaster 0.06 mm to 0.11 mm, about half maximum hind tibial width. Erect hairs on genae, scapes and tibiae numerous. Length: 3.8–5.5 mm.

Queen. Reddish brown. Head broader than maximum width of alitrunk. Funiculus segments longer than broad. Petiole sides curved, tapering to dorsal crest which is more or less emarginate. Pubescence and body hairs as in worker. Head width: 1.65–1.80 mm. Length: 6.8–8.0 mm.

Male. Dark brown to brownish black. Head broad with denticulate mandibles. Petiole somewhat tapering. Body surface including frons with rugose microsculpture and generally thick pubescence. Eyes with outstanding hairs. Tibial and scape hairs variable often sparse. Length: 4.0–4.8 mm.

Distribution. Local in Denmark and Southern Fennoscandia up to 62°. Throughout British Isles to Central Scotland. – Range: throughout Europe, widely distributed and not uncommon.

Biology. This species nests under boulders, in tree stumps and at the base of old trees. Workers are subterranean and seldom or never seen above ground. Flight period from mid August to late September. Single queens found colonies by invasion of and adoption in *Lasius niger, L. alienus* or occasionally *L. brunneus* nests. In late summer dealate queens often wander over the surface of *L. niger* nests, sometimes carrying a dead *L. niger* worker as a prelude to securing adoption.

Lasius affinis (Schenck, 1852:62)

This is an uncommon Central European species which has occurred in North Germany and Poland and also Belgium and the Netherlands. The worker is like that of *L. bicornis,* but has the dorsum of the gaster as well as alitrunk crowded with long hairs and the scale with a more angular emargination. Queens and males have similar differences and are much larger than *L. bicornis.*

Lasius distinguendus (Emery, 1916:64)

This is an inadequately characterised species. It is mainly found in South and Central Europe and most easily recognised in the queen caste. It is like a larger, paler *L. mixtus* but has more abundant genal hairs and a high broadly emarginate scale. The worker has occasional tibial hairs and longer body hairs than *L. mixtus* and would be more easily confused with *L. umbratus.* According to B. Pisarski (priv.commun.) it occurs in North Germany and probably also in Poland.

39. ***Lasius meridionalis*** (Bondroit, 1919)
Figs. 143–146.

Formica meridionalis Bondroit, 1919:143.

Lasius meridionalis (Bondroit); Pisarski, 1975.
Lasius rabaudi Bondroit, sensu Wilson, 1955; Collingwood, 1963; Kutter, 1977.

Worker: Clear yellow; pubescence on head rather dilute but close and very fine on gaster. Funiculus segments distinctly longer than wide; scapes and tibiae elliptical in cross section with thin front edge. Petiole sides straight to weakly convex, dorsal margin flat to slightly emarginate. Body and appendage hairs numerous. Length: 3.5–5 mm.

Queen. Dark brownish black; general apparance shining with fine shallow microsculpture. Body pubescence dilute but close and very fine on gaster. Funiculus segments longer than wide; scapes and tibiae flattened with thin front edge. Scale straight sided, dorsal margin flat, occasionally weakly emarginate. Body and appendage hairs abundant. Head width 1.7–1.8 mm. Length: 7.0–8.0 mm.

Male. Black; clypeus and frons distinctly shining with weak microsculpture. Pubescence sparse except on gaster where it is very fine and close. Frontal triangle,

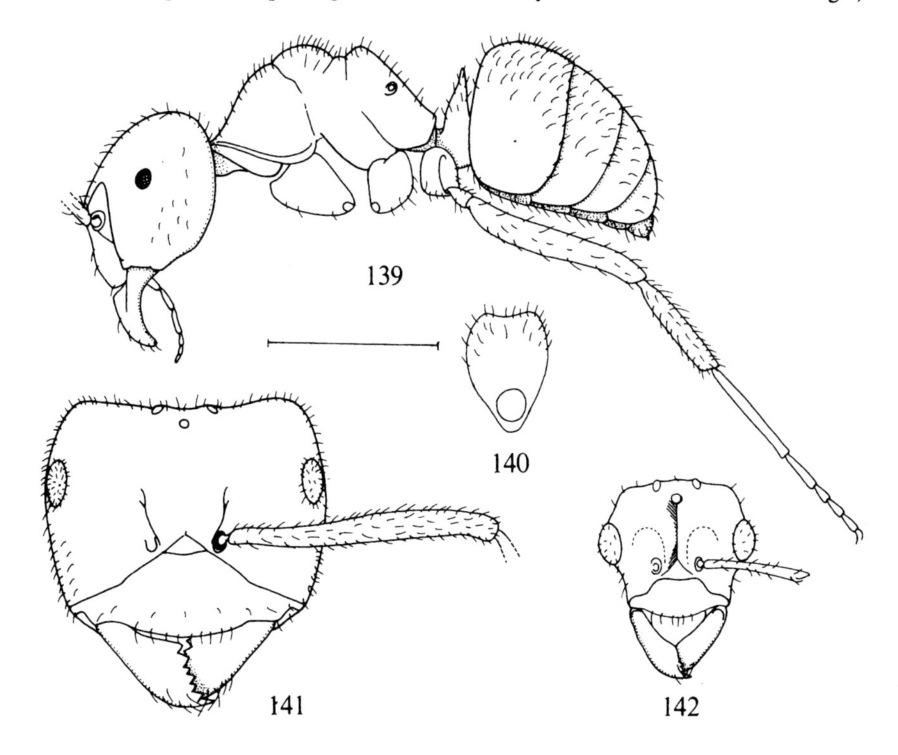

Figs. 139–142. *Lasius umbratus* (Nyl.). – 139: worker in profile; 140: petiole scale of queen in anterior view; 141: head of queen in dorsal view; 142: head of male in dorsal view. Scale: 1 mm.

frontal furrow and mandibular teeth very distinct. Head at least as broad as alitrunk. Eyes with erect hairs, appendage and body hairs numerous. Cross vein m-cu often absent on fore-wings. Length: 4.0–4.5 mm.

Distribution. Local; Denmark: EJ, NWJ, LFM, NEZ, B. – Sweden: Sk., Bl., Hall., Öl. and Dlr. – Norway: VE (Stolpestad). – Finland: N (Korverhar). – Locally common in Southeast England and South Wales. – Range: Spain to Japan, Italy to Scandinavia.

Biology. This species is characteristic of lowland sandy heath in North Europe. Nests are in the ground, often with low earth mounds and carton lined chambers. Flight period August. Fertilised queens start colonies through adoption by *L. alienus*. Males which have well toothed mandibles have been seen to pick up objects and to feed themselves.

Note. I have followed Pisarski (1975) in separating this species from *L. rabaudi*. According to examples of all castes kindly sent by P. Werner from Czechoslovakia, *L. rabaudi* has much more dilute but longer pubescence and has the gaster brilliantly shin-

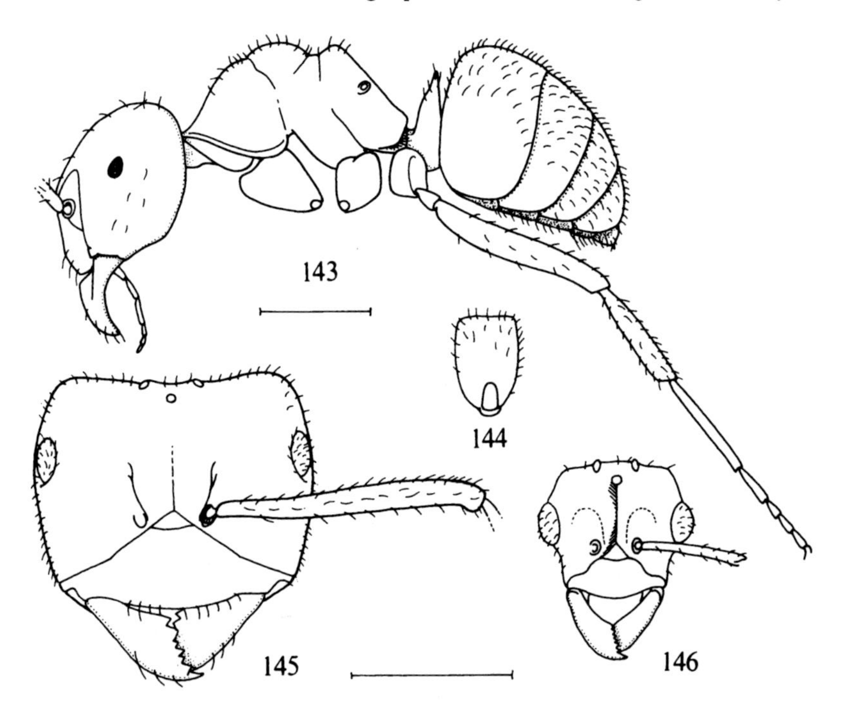

Figs. 143–146. *Lasius meridionalis* (Bondr.). – 143: worker in profile; 144: petiole scale of queen in anterior view; 145: head of queen in dorsal view; 146: head of male in dorsal view. Scale: 1 mm.

ing. Bourne (1973) synonymised *L. rabaudi* i. e. *L. meridionalis*, in England as *L. umbratus*. However, the different habits, flat appendages and rectangular scale in the queen, shining darker colour and fine sculpture in queen and male clearly distinguish the species from *L. umbratus*, although workers may be less easy to separate.

40. ***Lasius bicornis*** (Förster, 1850)
Figs. 147, 148.

Formica bicornis Förster, 1850:41.

Worker. Clear citron yellow. Funiculus segments not longer than broad; scape distinctly flattened. Outline of petiole characteristic: high, tapering dorsally with a deep emargination. Body hairs long, longest hairs nearly as long as maximum hind tibial width, sparse on dorsum of gaster where restricted to posterior borders of tergites; genal hairs sparse, scapes and tibiae bare. Length: 4.0–4.5 mm.

Queen. Yellowish brown to brown with head wider than alitrunk. Characters as worker; size relatively small. Length: 4.8–5.3 mm.

Male. Dark brown; petiole high, tapered and deeply indented; gaster hairs very sparse. Length: 4.0 mm.

Distribution. Rare; Sweden: Sm. (Forsslund, 1957a). – Range: Central and South Europe from Pyrenees and Caucasus but also recorded from Himalayas (Kashmir), South Italy to Netherlands; uncommon.

Biology. This species nests in rotten logs. Alatae have been taken in September (Poldi, 1962).

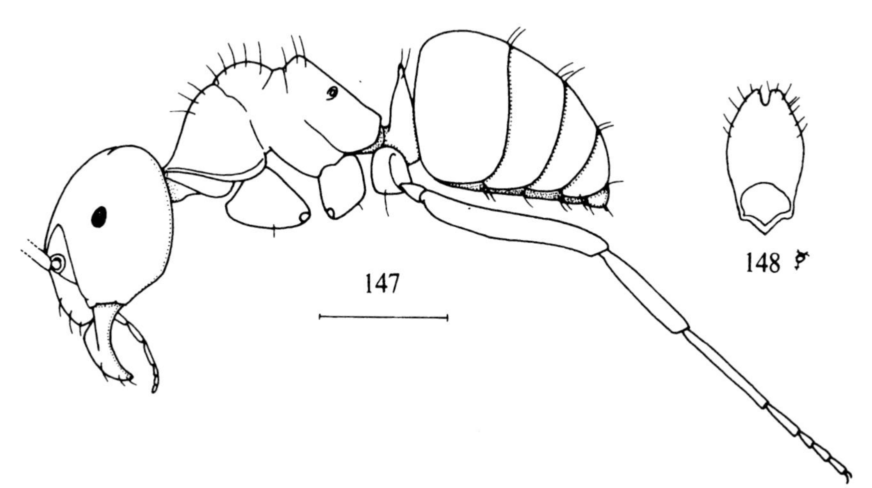

Figs. 147, 148. *Lasius bicornis* (Förster). – 147: worker in profile; 148: petiole scale of worker in anterior view. Scale: 1 mm.

41. ***Lasius mixtus*** (Nylander, 1846)
Figs. 149–152.

Formica mixta Nylander, 1846b:1050.

Worker. Yellow to brownish yellow; funiculus segments 2 to 5 not longer than broad; scape broadly oval in cross section. Petiole low, narrowing to emarginate dorsal border. Body pubescence coarse but sparse. Body hairs short, longest gaster hairs less than one third maximum hind tibial width. Front tibiae and scape without erect hairs, occasionally present on hind tibiae. Length: 3.5–4.5 mm.

Queen. Brownish black; head about as broad as maximum width of alitrunk. Funiculus segments 2 to 6 as broad as long. Scape broadly oval in cross section. Sides of petiole convex with emarginate dorsal border. Body pubescence sparse; microsculpture shallow but coarse. Body hairs short, rather sparse. Front tibiae and

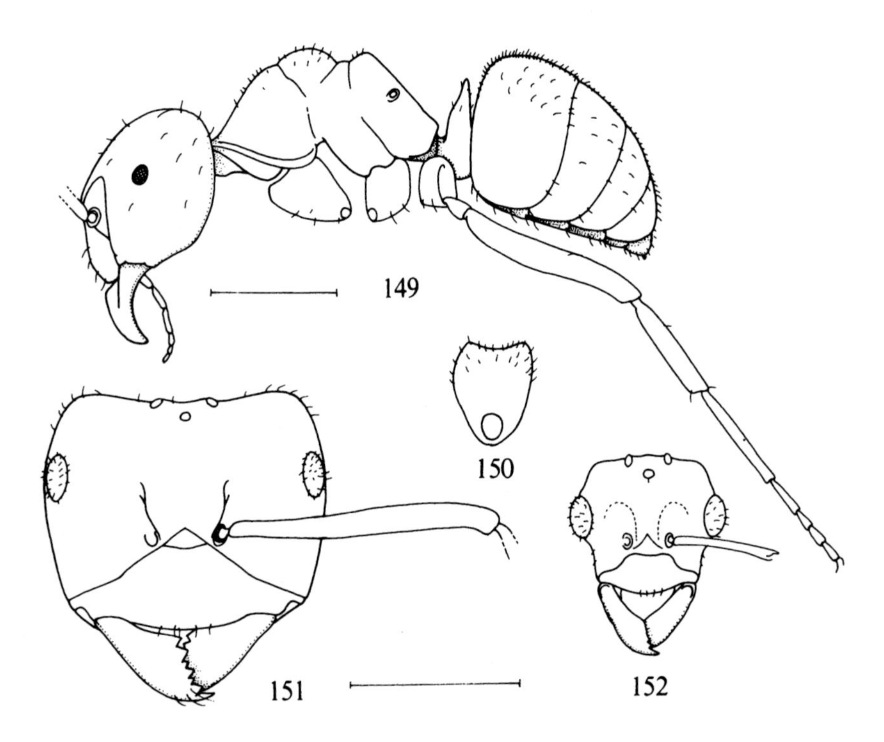

Figs. 149–152. *Lasius mixtus* (Nyl.). – 149: worker in profile; 150: petiole scale of queen in anterior view; 151: head of queen in dorsal view; 152: head of male in dorsal view. Scale: 1 mm.

scapes without suberect hairs, occasionally one or two present on hind tibiae. Eyes with short hairs between facets. Head width: 1.42–1.68 mm. Length: 6.0–7.5 mm.

Male. Dull brownish black; mandibular dentition weak and indistinct. Pubescence sparse, microsculpture coarse with clypeus and frons weakly rugose. Body hairs short, scape and tibiae without suberect hairs. Eyes indistinctly haired. Length: 4.2–4.8 mm.

Distribution. Throughout Denmark and South Fennoscandia to 62°. – Throughout the British Isles. – Range: throughout north Eurasia and subalpine regions of Central Europe.

Biology. This species nests deep in the ground often among shrub roots and under deep boulders, but occasionally also constructs mound nests of fine loose soil. Alatae fly in August and September and fertilised queens are thought to found fresh colonies through adoption by *Lasius alienus* and *L. niger* but actual recorded instances are very few or dubious. Dealate queens are often found wandering singly above ground in spring.

Note. *L. mixtus* was synonymised by Wilson (1955) as one extreme of a variable species *L. umbratus* (Nyl.). However in North Europe the characteristic *L. umbratus* with setose appendages is consistently distinct and justifies species separation (Collingwood, 1963b).

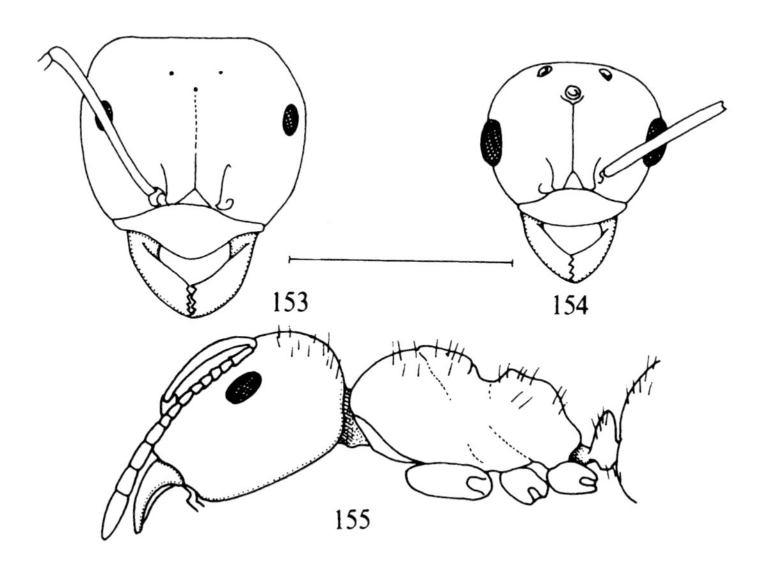

Figs. 153–155. *Lasius carniolicus* (Mayr). – 153: head of worker in dorsal view; 154: head of male in dorsal view; 155: head and alitrunk of worker in profile. Scale: 1 mm.

42. ***Lasius carniolicus*** (Mayr, 1861)
Figs. 153–155.

Lasius carniolicus Mayr, 1861:51.

Worker. Yellow to reddish yellow. Funiculus segments broad; scape oval in section; maxillary palps with segments 5 and 6 subequal, distinctly shorter than 4. Petiole in profile nodal with both anterior and posterior faces convex; in front view with rounded dorsal surface. Sides of head curving forward into close set mandibles; head cordate with rounded occipital corners, posterior margin slightly concave. Body pubescence long and thick, merging into short suberect hairs on all surfaces. Length: 3.5–3.7 mm.

Queen. Yellowish brown. Petiole as in worker. Pubescence long. Eyes haired. Mandibles with concave masticatory border 5 toothed, with prominent long narrow apical tooth; wings hyaline. Size small – head width: 0.76–0.78 mm, broader than alitrunk. Length: 3.6–3.7 mm.

Male. Blackish brown. Petiole low and convex in lateral view, dorsal margin flat in frontal view. Wings hyaline, vein m-cu missing. Body hairs abundant, scapes and tibiae with decumbent pubescence only. Head as broad as alitrunk, mandibles dentate. Length: 3.3–3.5 mm.

Distribution. Sweden: Öl. and Gtl., recently also recorded from Småland on the mainland (Douwes, 1976). Very rare. – Range: Pyrenees to Siberia, Italy to Poland and South Sweden, also Afghanistan, Karakorum and Himalayas (Faber, 1967).

Biology. This species has a wide but scattered occurrence, rather local but sometimes abundant in the restricted areas where it occurs. Nests are found under stones in sandy soil in open dry woodland or pasture. The queens which are no larger than the worker start colonies by adoption in nests of *Lasius alienus* or *L. flavus*. Flight period September/October. Faber (1967) records extreme physogastry in the mature egg laying queen of the related species *L. reginae* Faber.

Genus *Paratrechina* Motschulsky, 1863

Paratrechina Motschulsky, 1863: 13.
Type-species: *Formica longicornis* Latreille, 1802.

Mandibles narrow; palp formula 6, 4. Antennae 12 segmented in female castes, 13 segmented in males. Eyes set at or in front of mid length of head. Ocelli absent in workers. Propodeum and scale unarmed. Petiole scale reduced inclined forward, in many species overhung by first gaster segment. Dorsal surface of head, alitrunk and gaster with paired coarse setae.

A pantropical genus with many species including a few in temperate Asia and several cosmopolitan species of which two are frequently introduced on plant material into Europe and become established in heated premises.

Keys to species of *Paratrechina*

Workers

1 Alitrunk conspicuously elongated. Antennae very long, extending backward as far as propodeum. Appendage setae long and sparse, usually absent on scapes. Length: 2.5–3.5 mm. Colour brownish black (Fig. 156) .... *longicornis* (Latreille)
– Alitrunk not conspicuously elongate. Antennal scape extending only as far back as mesonotum. Setae short and crowded on all appendages including scapes. Length: 2.0–3.0 mm. Colour yellowish brown to black (Fig. 157) .... *vividula* (Nylander)

Queens

1 Antennal scape over-reaching occipital border by over half its length, without outstanding hairs. Length: 5.0–5.5 mm .... *longicornis* (Latreille)
– Antennal scape not exceeding occipital border, with numerous erect hairs. Length: 3.5–5.0 mm .... *vividula* (Nylander)

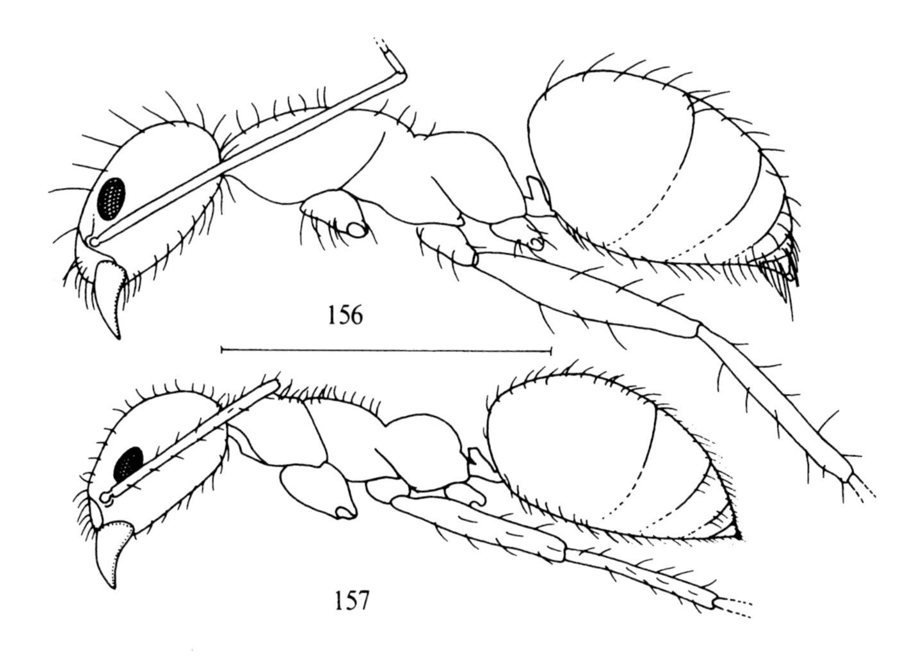

Fig. 156. *Paratrechina longicornis* (Latr.), worker in profile. Scale: 1 mm.
Fig. 157. *Paratrechina vividula* (Nyl.), worker in profile. Scale: 1 mm.

Males

1 Head longer than broad; scape without hairs. Length: 2.5 mm *longicornis* (Latreille)
– Head not longer than broad; scape with outstanding hairs. Length: 2.0 mm *vividula* (Nylander)

Paratrechina longicornis (Latr.) (Fig. 156) is of frequent occurrence in the British Isles. Established colonies may be very populous with many queens and are difficult to eradicate.

Paratrechina vividula (Nyl.) (Fig. 157) has been recorded from heated glasshouses in botanic gardens in Denmark, Uppsala and Göteborg in Sweden and Helsinki in Finland as well as many places in the British Isles.

TRIBE PLAGIOLEPIDINI FOREL

Genus *Plagiolepis* Mayr, 1861

Plagiolepis Mayr, 1861: 42.
Type-species: *Formica pygmaea* Latreille, 1798.

This genus includes several European species, none of which have occurred in the British Isles, Denmark or Fennoscandia. However one species occurs in the Channel Islands and has been recorded from Belgium and Poland so is briefly described here for completeness.

Plagiolepis vindobonensis Lomnicki, 1925.
Fig. 158.

Plagiolepis vindobonenis Lomnicki, 1925: 78.

Worker. Pale to dark brown, somewhat shining, with sparse pubescence. Erect hairs are present over head and alitrunk and more thickly distributed over gaster. Head about as long as broad; pronotum broad, about 1½ width of propodeum; gaster heartshaped, much broader than alitrunk; scale, nodal reduced. Third and fourth funiculus segments subequal, about twice as long as second segment. Scape overreaching occiput by about 1/6 its length. Ocelli absent. Length: 1.0–2.0 mm.

Queen. As worker but with broad flat alitrunk. Length: 3.0–4.0 mm.

Male. Brownish black. Head broader than long; eyes prominent, set forward of midlength of head. Antennae 12 segmented with scape overreaching occiput by 1/5 its length, funiculus segments scarcely longer than broad. Length: 1.5–2.0 mm.

Distribution. Central and Eastern Europe.

Biology. This minute species lives in small isolated colonies nesting under flat stones usually with several queens.

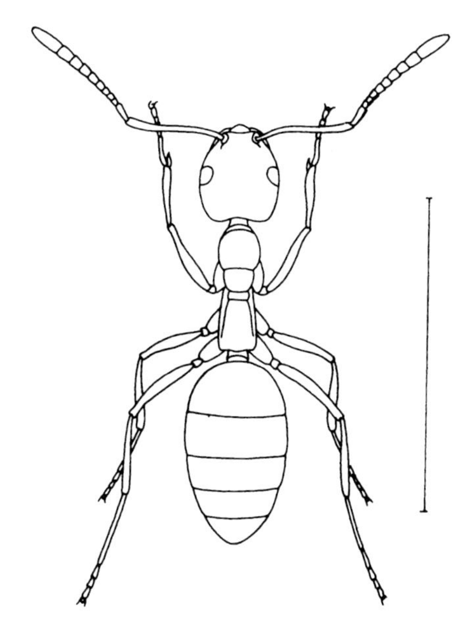

Fig. 158. *Plagiolepis vindobonensis* Lomnicki, worker in dorsal view. Scale: 1 mm.

TRIBE FORMICINI FOREL

Genus *Formica* Linné, 1758

Formica Linné, 1758: 579.
Type-species: *Formica rufa* Linné, 1758 (Yarrow, 1954).

Antennae 12 segmented in worker and female, 13 segmented in male; antennal insertions contiguous with clypeal border. Funiculus segments 2 to 5 longer than 6 to 10. Maxillary palps 6 segmented, occasionally 5; labial palps 4 segmented. Ocelli distinct in all castes; frontal ridges diverging posteriorly; frontal triangle always very distinctly defined. Propodeal spiracle elliptical placed at some distance from propodeal border. Wings with cubital and one discoidal cell. Male external genitalia large and conspicuous. The species of this genus are relatively robust and active.

Holarctic genus with over 150 species. There are at least 20 species known to occur in Fennoscandia; in some groups, notably *F. rufa* and its allies, species distinctions are not always clearcut necessitating the comparison of a number of specimens preferably with males and females as well as workers.

Keys to species of *Formica*

Workers

1 Bicoloured species: alitrunk red with varying amounts of dark brown to blackish patches 2

– Body colour evenly greyish or brownish black 18

2 (1) Anterior margin of clypeus emarginate or with distinct median notch (Fig. 222) .... 56. *sanguinea* Latreille

– Anterior margin of clypeus broadly rounded and entire 3

3 (2) Head flattened posteriorly, with deeply concave occipital border 4

– Head evenly rounded posteriorly, occipital margin straight, convex or very weakly concave 8

4 (3) Eyes with distinct microscopic hairs; maxillary palps longer than half head length (Figs. 201, 212) .... 50. *exsecta* Nylander

– Eyes bare; maxillary palps half head length or shorter 5

5 (4) Occipital corners smoothly rounded; posterior emargination shallow, head not longer than maximum width across eyes; maxillary palps 6 segmented, as long as half head length (Fig. 213) .... 54. *suecica* Adlerz

– Occipital corners sharp; back of head deeply emarginate; maxillary palps often 5 segmented, shorter than half head length; head longer than broad 6

6 (5) Dorsal hairs on gaster few but present on all tergites; middle of clypeus as well as its anterior border with projecting hairs (Figs. 204, 205) 52. *forsslundi* Lohmander

– Dorsal hairs on gaster restricted to posterior margin of third tergite to apex; clypeal hairs restricted to anterior border (Fig. 207) 7

7 (6) Gaster moderately shining, with dilute sculpture and pubescent hairs shorter than their interspaces. Clypeus transversely impressed below midline (Figs. 206, 215) .... 53. *pressilabris* Nylander

– Basal tergite of gaster with upper surface sculptured and dull; pubescent hairs slightly longer than their interspaces. Clypeus with or without shallow impression below midline (Fig. 216) .... 51. *foreli* Emery

8 (3) Frontal triangle dull; terminal segment of maxillary palp as long as fifth segment 9

– Frontal triangle reflecting light; terminal segment of maxillary palp shorter than fifth segment 12

9 (8) Head and frontal triangle coarsely sculptured, opaque black. Head as broad as long; antennal scape short, not over-reaching occipital margin by more than 1/4 its length (Fig. 219) .... 55. *uralensis* Ruzsky

– At least genal margins and clypeus reddish, sculpture weaker but pubescense strongly developed and moderately thick especially on gaster. Head distinctly longer than broad; antennal scape slender, over-reaching occipital margin by 1/3 its length or more 10

10 (9) Whole body including gula and posterior margin of head with numerous projecting hairs (Figs. 162, 188) .... 47. *cinerea* Mayr

– Gula and posterior margin of head entirely without hairs 11

11 (10) Dorsum of promesonotum and upper margin of petiole scale with projecting hairs (Fig. 197) .... 49. *rufibarbis* Fabricius

– Promesonotum and scale without projecting hairs, occasionally one or two short hairs present on promesonotum (Fig. 193) .... 48. *cunicularia* Latreille

12 (8) Antennae long with scapes longer than head width; second and third funiculus segments in larger workers twice as long as broad (Fig. 225)
57. *truncorum* Fabricius

– Antennal scape shorter than head width; funiculus segments always less than twice as long as broad 13

13 (12) Gaster thickly pubescent and dull; frons closely sculptured and dull; dark patch on promesonotum normally clearly defined and opaque 14

– Gaster and frons moderately shining; dark patch on promesonotum very variable, not normally clearly defined and sometimes absent 15

14 (13) Head and body profusely hairy with longest hairs on alitrunk × 0.1 or more head width. Antennal scapes usually with two or three short projecting subdecumbent hairs (Fig. 266) .... 63. *nigricans* Emery

– Pilosity variable, alitrunk hairs usually less than × 0.1 head width. Antennal scapes with adpressed pubescence only (Figs. 256, 261) .... 62. *pratensis* Retzius

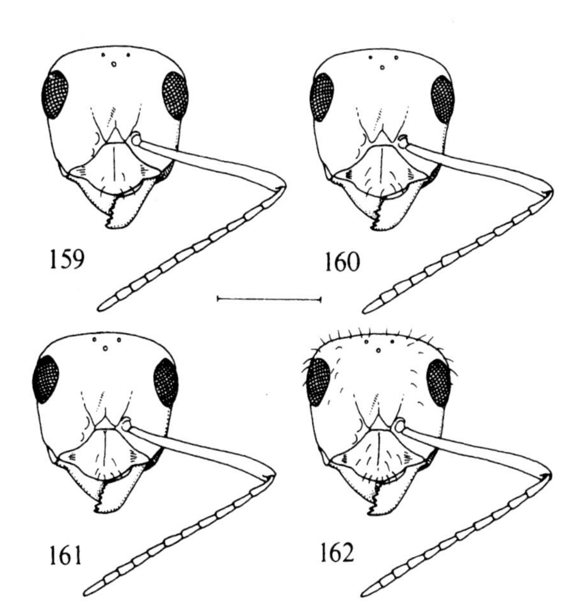

Figs. 159–162. Head of *Formica*-workers in dorsal view. – 159: *fusca* (L.); 160: *rufibarbis* Fabr.; 161: *cunicularia* Latr.; 162: *cinerea* Mayr. Scale: 1 mm.

15 (13) Eyes, gula, alitrunk and posterior margin of head conspicuously hairy, with hairs extending forward laterally as far as eyes (Figs. 249, 254) .... 61. *lugubris* Zetterstedt
– Eyes with very short hairs or bare. Fringe of hairs at back of head when present not extending forward beyond occipital corners 16
16 (15) First tergite of gaster evenly and closely set with micropunctures; eyes always with distinct short hairs; frons closely sculptured and rather dull; extensor surface of hind femora and tibiae always with some projecting hairs; posterior of head with projecting hairs, sometimes very few or absent (Figs. 242, 247) .... 60. *aquilonia* Yarrow
– First tergite of gaster with punctures widely spaced on upper medial area; eyes bare or with very few microscopic hairs; frons moderately shining with shallow sculpture and easily seen scattered coarse punctures. Back of head without projecting hairs 17
17 (16) Gula, dorsum of head and alitrunk with conspicuous projecting hairs; occasional or few projecting hairs normally present on extensor surface of hind femora and tibiae. Mesopleurae from above with evenly distributed long hairs (Fig. 229) .... 58. *rufa* Linné
– Gula bare or with one or two weak hairs, dorsum of head and alitrunk with occasional short hairs. Hind femora and tibiae normally bare. From above projecting hairs on mesopleurae restricted to a few only at posterior end (Fig. 236) .... 59. *polyctena* Förster
18 (1) Back and underside of head with copious erect hairs (Fig. 188) .... 47. *cinerea* Mayr
– Back of head bare 19
19 (18) Dorsum of promesonotum with erect hairs 20
– Dorsum of promesonotum bare or with occasional one or two hairs only 21
20 (19) Whole body brilliantly shining; promesonotal hairs pointed, long and curving forward; one or two gula hairs often present (Fig. 184) .... 46. *transkaucasica* Nasonov
– Body with subdued shine; promesonotal bristles short and blunt; gula hairs never present (Fig. 182) .... 45. *lemani* Bondroit
21 (19) Gaster shining, with very dilute pubescence and absence of sculpture medially; petiole scale emarginate dorsally .... 44. *gagatoides* Ruzsky
– Gaster with close pubescence and microsculpture; petiole scale with rounded or flat dorsal crest .... 43. *fusca* Linné

Females

1 Anterior margin of clypeus impressed or notched .... 56. *sanguinea* Latreille
– Anterior margin of clypeus entire, broadly convex 2
2 (1) Occipital margin of head deeply excised or broadly emarginate 3
– Occipital margin of head convex 7

3 (2) Eyes with copious hairs; size larger, length: 7.5–9.0 mm ... 50. *exsecta* Nylander
– Eyes bare; size smaller, length: 5.5–7.2 mm ... 4
4 (3) Occipital corners smoothly rounded, emargination of head shallow; petiole scale with angular lobes; maxillary palps 6 segmented .... 54. *suecica* Adlerz
– Occipital corners sharp, enclosing deep emargination; petiole scale with rounded lobes; maxillary palps short, normally 5 segmented ... 5
5 (4) Erect hairs present on dorsum of second gaster tergite to apex; middle of clypeus with long hairs ... 52. *forsslundi* Lohmander
– Erect hairs on dorsum of gaster restricted to posterior border of third tergite to apex; clypeal hairs restricted to anterior border only ... 6
6 (5) Body uniformly shining dark brown ... 53. *pressilabris* Nylander
– Gaster somewhat matt appearance, propodeum and petiole yellowish red 51. *foreli* Emery
7 (2) Body colour uniformly brownish black or black ... 8
– Body bicoloured with portions of alitrunk at least distinctly reddish ... 12
8 (7) Back of head and gula with copious erect hairs (Fig. 190) ...... 47. *cinerea* Mayr
– Back of head always bare ... 9
9 (8) Underside of mid femora bare; in dorsal view hairs on pronotum restricted to anterior portion only ... 10
– Underside of mid femora with several long hairs; erect hairs on pronotum extend round the sides to the tegulae ... 11
10 (9) Gaster and scutellum brilliant; middle of gaster dorsum impunctate, pubescence very sparse and dilute; petiole scale broadly heart-shaped; gula hairs occasionally present (Fig. 181) ... 44. *gagatoides* Ruzsky
– Gaster and scutellum with subdued shine, with distinct microsculpture and pubescence; petiole scale normally with dorsal crest flat or convex; gula hairs never present (Fig. 178) ... 43. *fusca* Linné
11 (9) Whole body shining; pubescence on gaster long but sparse; 2 or 3 gula hairs normally present (Fig. 187) ... 46. *transkaucasica* Nasonov
– Gaster dull with close pubescence and distinct microsculpture; underside of head always bare (Fig. 183) ... 45. *lemani* Bondroit
12 (7) Frontal triangle densely sculptured or dull. Eyes completely hairless. Terminal segment of maxillary palp as long as fifth ... 13
– Frontal triangle reflecting light. Eyes with conspicuous hairs or with occasional microscopic hairs. Terminal segment of maxillary palp shorter than preceding segments ... 16
13 (12) Whole body including gula and posterior margin of head with profuse hairs; body colour mainly brownish black, but occasionally genae and mesopleural articulations reddish (Fig. 190) ... 47. *cinerea* Mayr
– Hairs entirely absent on posterior margin of head ... 14
14 (13) Head entirely black, coarsely sculptured; antennal scape short and thick; femora black; underside of head usually with a few coarse hairs 55. *uralensis* Ruzsky

– Genae and clypeus distinctly reddish; antennal scapes slender; femora yellowish red; underside of head always bare 15

15 (14) Propodeum with a few erect hairs; mesoscutum in part reddish (Fig. 199)
49. *rufibarbis* Fabricius

– Propodeum without erect hairs; mesoscutum normally entirely dark (Fig. 195) .... 48. *cunicularia* Latreille

16 (12) Gaster and whole of scutum matt with more or less dense pubescence 17

– Gaster and scutellum distinctly shining; pubescence dilute, not obscuring cuticular shine 19

17 (16) Second funiculus segment twice as long as broad. Whole body including mesonotum with large shallow punctures. Head normally with red colour predominating (Fig. 226) .... 57. *truncorum* Fabricius

– Second funiculus segment always less than twice as long as broad. Larger punctures deep and normally concealed by pubescence 18

18 (17) Whole body including propodeum, scale and basal face of gaster with conspicuous long bent hairs. Antennal scape often with two or three projecting hairs (Figs. 263, 264) .... 63. *nigricans* Emery

– Projecting hairs on posterior margin of head and basal face of gaster normally short, absent on dorsum of propodeum and scale. Scapes without projecting hairs (Figs. 258, 259) .... 62. *pratensis* Retzius

19 (16) Back of head, basal face of gaster scale and propodeum with long hairs (Figs. 251, 252) .... 61. *lugubris* Zetterstedt

– Dorsum of scale and propodeum without hairs, back of head and basal face of gaster either bare or with short sparse hairs 20

20 (19) Gaster including medial dorsal area of first tergite very finely and closely set with micropunctures. Eyes with distinct short erect hairs (Figs. 174, 244, 245) .... 60. *aquilonia* Yarrow

– Micropunctures on gaster sparse and widely set, larger punctures well spaced and clearly visible. Eyes with a few microscopic hairs or bare (Figs. 171, 172, 232, 239) 21

21 (20) Middle of scutellum normally brilliant, without punctures or striae.
58. *rufa* Linné

– Middle of scutellum with longitudinal striae and punctures rather dull
59. *polyctena* Förster

Males

1 Back of head broadly excavate with pronounced occipital angles 2

– Back of head convex 6

2 (1) Occipital margins and eyes conspicuously hairy (Fig. 208)
50. *exsecta* Nylander

– Occipital margins and eyes bare or with very sparse short hairs 3

3 (2) Mesonotum with scattered hairs; maxillary palps 6 segmented; occipital corners smoothly rounded (Fig. 209) .... 54. *suecica* Adlerz

– Mesonotum bare; maxillary palps short, normally 5 segmented; occipital corners sharp 4

4 (3) Outstanding hairs on dorsum of gaster present on second tergite to apex 52. *forsslundi* Lohmander

– Outstanding hairs on gaster restricted to fifth tergite to apex 5

5 (4) Body shining; pubescent hairs on gaster shorter than interspace; eyes with 1 or 2 microscopic hairs at most .... 53. *pressilabris* Nylander

– Body somewhat dull; pubescent hairs on gaster longer than interspaces; eyes with scattered microscopic hairs .... 51. *foreli* Emery

6 (1) Front border of clypeus emarginate or notched in middle (Fig. 221) 56. *sanguinea* Latreille

– Front border of clypeus not impressed, with straight or convex front border 7

7 (6) Eyes with conspicuous hairs 8

– Eyes bare 14

8 (7) Whole body thickly haired, in side view suberect hairs on dorsum of gaster numerous and uninterrupted from base to apex. Frontal triangle without punctures or coarse sculpture. Legs and external genitalia mainly yellow. Antennae long and slender, second and third funiculus segment more than × 2.5 as long as broad (Fig. 227) .... 57. *truncorum* Fabricius

– In side view hairs on dorsum of gaster sparse and not appearing as an uninterrupted fringe. Frontal triangle shining but normally micropunctures and/or some pubescence present. Genitalia and legs with at least femora partly infuscated. Funiculus segments less than × 2.5 as long as broad 9

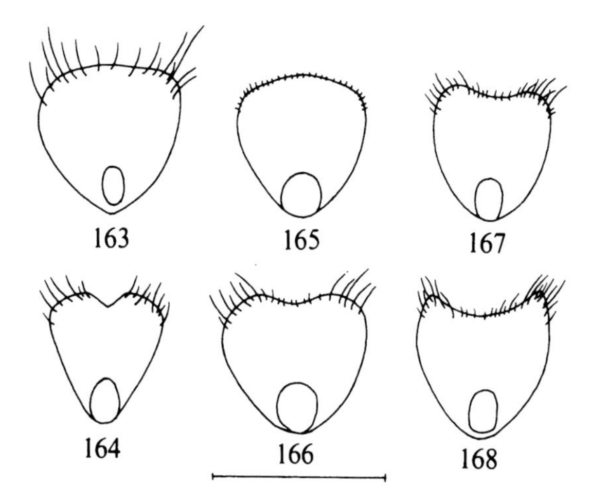

Figs. 163–168. Petiole scale of *Formica*-males in posterior view. – 163: *transkaucasica* Nasonov; 164: *gagatoides* Ruzsky; 165: *fusca* (L.); 166: *lemani* Bondr., 167: *cunicularia* Latr.; 168: *rufibarbis* Fabr. Scale: 1 mm.

9 (8) Scutellum and gaster pubescent and dull (eyes and genae always with profuse hairs) 10

– Scutellum and gaster at least moderately shining 11

10 (9) Tibial hairs numerous, with occasional long erect hairs on hind tibiae at least × 0.5 maximum tibial width .... 63. *nigricans* Emery

– Tibial hairs short often sparse, none exceeding × 0.3 hind tibial width (Fig. 257) .... 62. *pratensis* Retzius

11 (9) Genal hairs profuse. Tibial hairs long and profuse (Figs. 250, 253) 61. *lugubris* Zetterstedt

– Genal margins below eyes either with a few projecting hairs or none. Tibial hairs short and sparse or absent 12

12 (11) Frons and gaster very finely punctured. Fringe of short hairs always present on upper surface of hind femora. In normal populations projecting hairs are present on the genae below eyes (Fig. 246) .... 60. *aquilonia* Yarrow

– Frons and gaster widely and coarsely punctured. Upper surface of hind femora with occasional hairs or none 13

13 (12) Metanotum bare or with sparse hairs. Longer hairs on mesonotum sparse. Genae never with projecting hairs below eyes (Fig. 237) 59. *polyctena* Förster

– Metanotum with long hairs which are also numerous over mesonotum. Genal projecting hairs occasionally present in some populations, normally absent (Fig. 230) .... 58. *rufa* Linné

14 (7) Scutellum dull or opaque 15

– Scutellum moderately shining or brilliant 18

15 (14) Antennal scapes thick – width equal to more than half narrowest eye width; head broad and not conspicuously narrowed anteriorly; femora and whole body opaque black, coarsely sculptured, petiole scale broadly convex (Fig. 220) .... 55. *uralensis* Ruzsky

– Antennal scapes slender – maximum width less than × 0.3 minimum eye width; head narrowing conspicuously towards clypeal border; sculpture fine but obscured by more or less thick pubescence on dorsal surfaces; legs yellowish or yellowish red with only femora sometimes infuscate, petiole scale emarginate 16

16 (15) Underside of head with erect hairs (Fig. 192) .... 47. *cinerea* Mayr

– Underside of head always bare 17

17 (16) Femora dark at least in part. Dorsal border of petiole scale sharply angulate at sides (Fig. 168) .... 49. *rufibarbis* Fabricius

---

Figs. 169, 170. Electroscan micrographs of *Formica*-eyes. – 169: worker of *rufa* L. 170: worker of *pratensis* Retzius. Scale: 100 μ.

Figs. 171–175. Electroscan micrographs of dorsum of first gaster tergite in *Formica*-queens. – 171: *rufa* L., 172: *polyctena* Förster; 173: *lugubris* Zett., 174: *aquilonia* Yarrow; 175: *pratensis* Retzius. Scale: 100 μ.

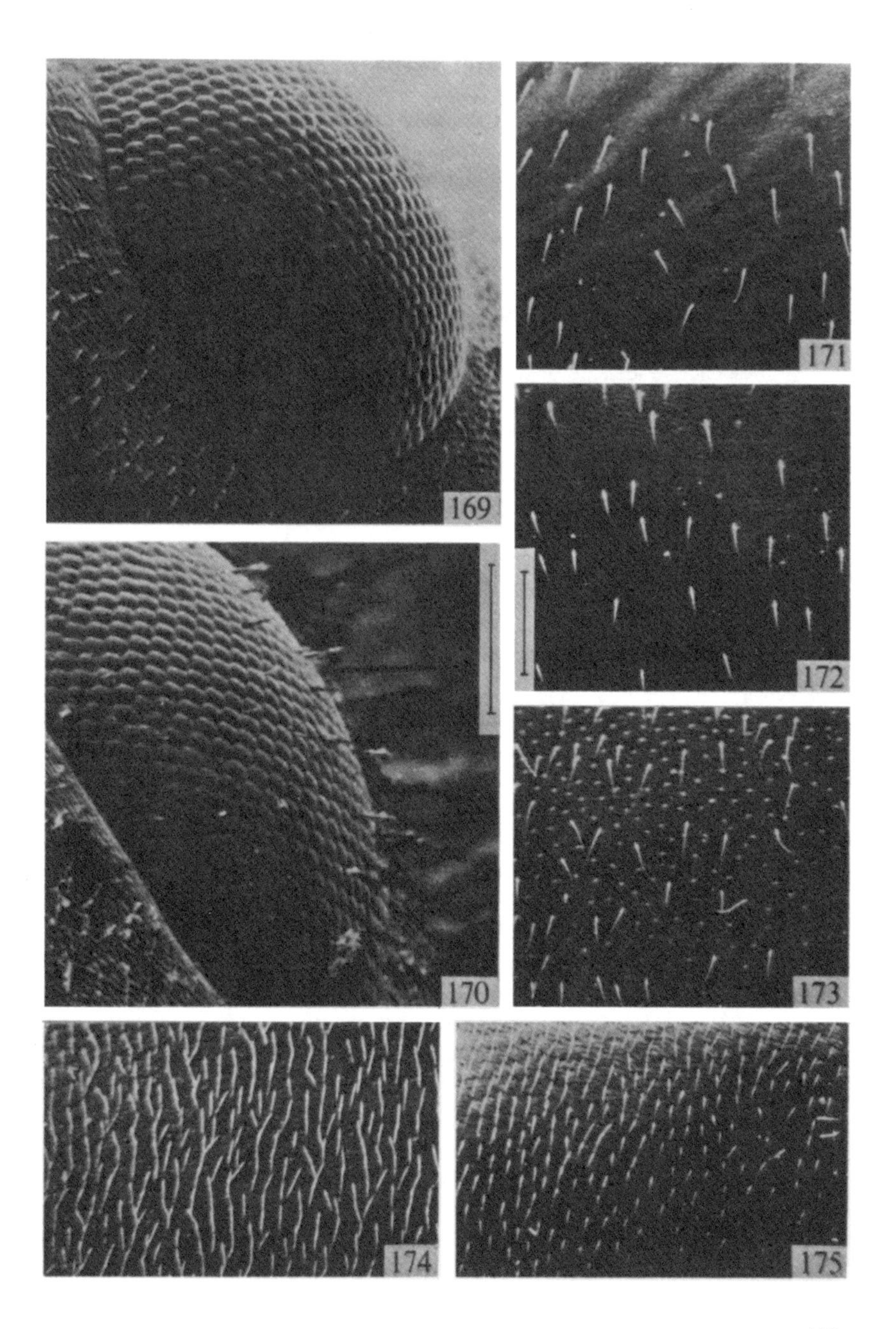
169
170
171
172
173
174
175

– Femora unicolorous with tibiae reddish or pale reddish yellow, sides of petiole scale rounded .... 48. *cunicularia* Latreille

18 (14) Petiole scale with very short fine hairs only; normally with rounded or flat dorsal border (Fig. 165) .... 43. *fusca* Linné

– Petiole scale with long hairs; dorsal border more or less deeply emarginate (Figs. 163, 164, 166) 19

19 (18) Underside of mid and hind femora bare; erect hairs on promesonotum sparse, short and weak .... 44. *gagatoides* Ruzsky

– Underside of mid femora with one or more scattered hairs, semi-erect hairs on promesonotum numerous and strong 20

20 (19) Gaster brilliant almost without sculpture medially, with long fine scattered pubescence; underside of head normally with one or two long hairs; frontal triangle shining .... 46. *transkaucasica* Nasonov

– Gaster moderately shining with short close pubescence and distinct microsculpture. Underside of head never with hairs; frontal triangle dull 45. *lemani* Bondroit

43. ***Formica fusca*** Linné, 1758
Figs. 159, 165, 176–179.

Formica fusca Linné, 1758: 580; Yarrow, 1954: 228 (redescription).

Worker. Black, legs brownish. Gula, occiput, mid femora and promesonotum without standing hairs – occasionally one or two weak pronotal hairs. Pubescent hairs on gaster longer than their interspace width. Frons with fine microsculpture. Length: 4.5–7.0 mm.

Queen. Colour and sculpture as worker. Pronotal hairs restricted to anterior part; underside of mid and hind femora without long hairs or those restricted to proximal part only. Scutellum sculptured, not conspicuously shining. Length: 7.0–9.5 mm.

Male. Black, appendages yellowish to brownish. Scale with dorsal fringe of very fine short hairs but no outstanding long hairs. Gaster with short adpressed pubescence. Length: 8.0–9.5 mm.

Distribution. Common throughout Denmark and Southern Fennoscandia to about latitude 63° in Finland. – Common in South England, Wales and Southwest Ireland, local in North England and Scotland. – Range: whole palaearctic region from Portugal to Japan, Italy to Central Fennoscandia.

Biology. This is the common black ant of Europe. It nests variously in banks, under stones and in tree stumps along hedgerows and woodland borders. Workers are timid, fast moving and forage singly, predating small insects but also feeding on extra floral nectaries and on aphid honeydew. Colonies are usually small with up to 500 workers and one or a few queens present. Alatae are developed in June and July and fly off the nests in July and early August.

44. ***Formica gagatoides*** Ruzsky, 1904
Figs. 164, 180, 181.

Formica fusca var. *gagatoides* Ruzsky, 1904: 377.
Formica gagatoides Ruzsky; Holgersen, 1943: 10 (redescription).

Worker. Black, mandibles and appendages brown. Head and alitrunk dull, gaster shining with sparse adpressed pubescent hairs – length less than half interspace width. Propodeum angled in profile. Petiole scale broadly heart-scaped with more or less emarginate dorsal border. Erect hairs on gaster restricted to posterior border of tergites; mid femora normally without outstanding hairs. Length: 4.2–6.0 mm.

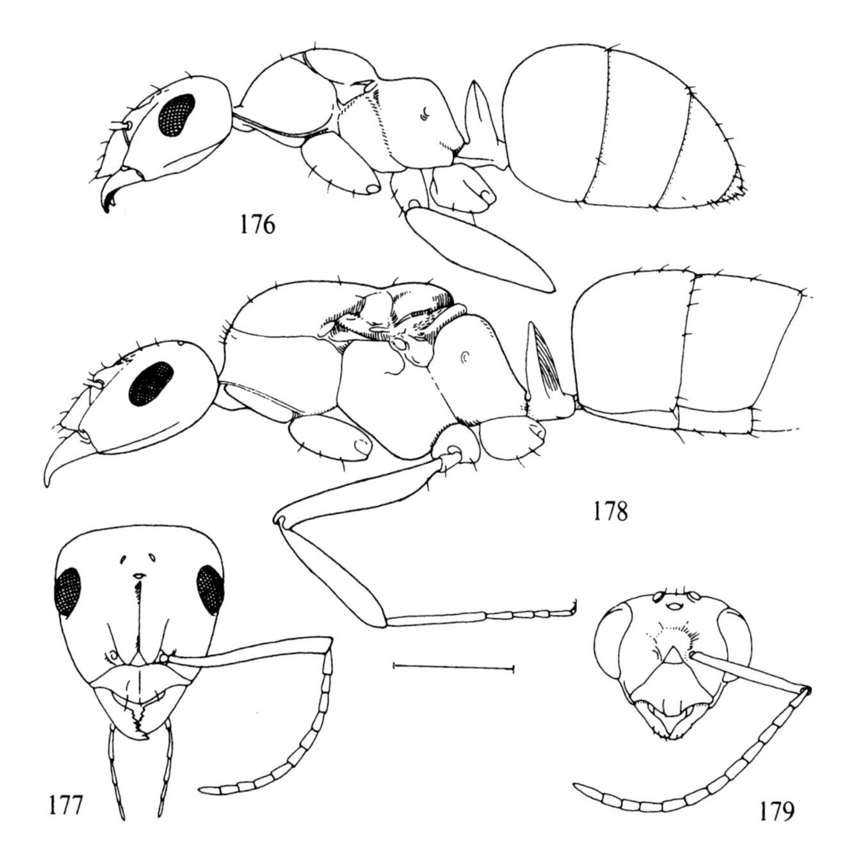

Figs. 176–179. *Formica fusca* (L.). – 176: worker in profile; 177: head of queen in dorsal view; 178: queen in profile; 179: head of male in dorsal view. Scale: 1 mm.

Queen. Colour and pubescence as worker. Scutellum and gaster conspicuously shining. Pronotal hairs sparse and restricted to anterior part; mid femora bare. Length: 7.0–8.0 mm.

Male. Black, legs and external genitalia yellowish. Head and alitrunk finely punctured with scutellum and propodeum as well as gaster distinctly shining. Scale broadest at apex, with shallow emargination and with scattered long hairs overreaching rounded side margins of dorsal crest. Length: 6.0–7.0 mm.

Distribution. North and Central Fennoscandia from North Cape south to Buskerud and Opland in Norway, Värmland and Medelpad in Sweden and Tavastia borealis in Finland. Abundant in northern areas and in mountains, more local in centre and absent south of latitude 60° in Norway to latitude 62° in Finland. – Range: exclusively arctic from Norway to Northeast Siberia.

Biology. This is one of the few Fennoscandian species that does not occur in the Alps or other mountains of Central Europe. In behaviour and general appearance it resembles *F. fusca,* which it replaces in the north, but can be immediately distinguished by the shining gaster from *F. fusca* and from *F. transkaucasica* by the duller head and alitrunk. It lives in small colonies of a few hundred workers with one or a few queens. Alatae fly in July and August.

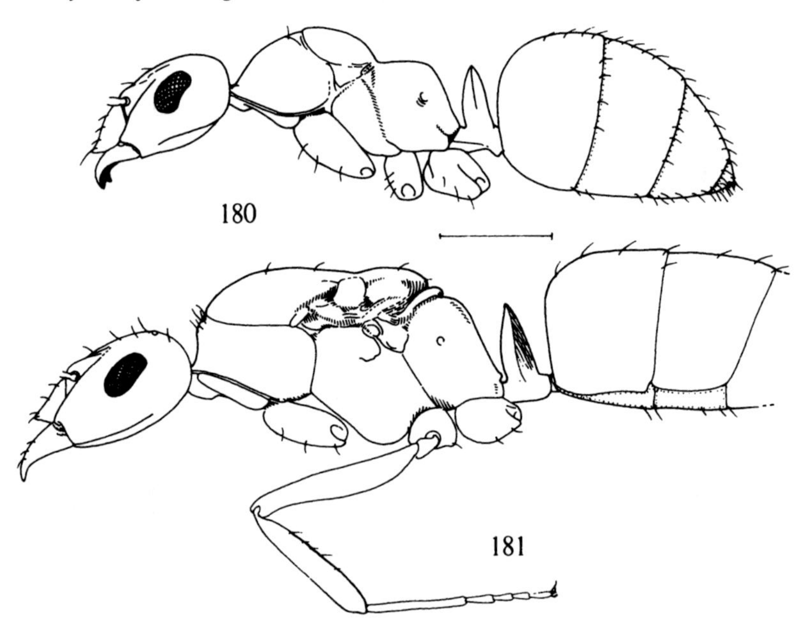

Figs. 180, 181. *Formica gagatoides* Ruzsky. – 180: worker in profile; 181: queen in profile. Scale: 1 mm.

45. ***Formica lemani*** Bondroit, 1917
Figs. 166, 182, 183.

Formica lemani Bondroit, 1917: 186; Yarrow, 1954: 230 (redescription).

Worker. Greyish to brownish black, legs paler. Short stout hairs present on promesonotum normally numerous but occasionally abraded or few. Underside of mid and hind femora normally with one or more hairs at mid length. Frons coarsely sculptured to that punctures readily seen under ordinary magnification. Length: 4.5–7.0 mm.

Queen. Colour, sculpture and pilosity as worker but scutellum shining and pronotal hairs numerous, extending round side margins to tegulae. Long hairs on underside of mid femora always present. Length: 7.0–9.5 mm.

Male. Black with appendages yellowish or brownish. Scale with conspicuous long hairs overreaching dorsal crest, most numerous at angulate side corners. Gaster with short adpressed pubescence. Length: 8.0–9.0 mm.

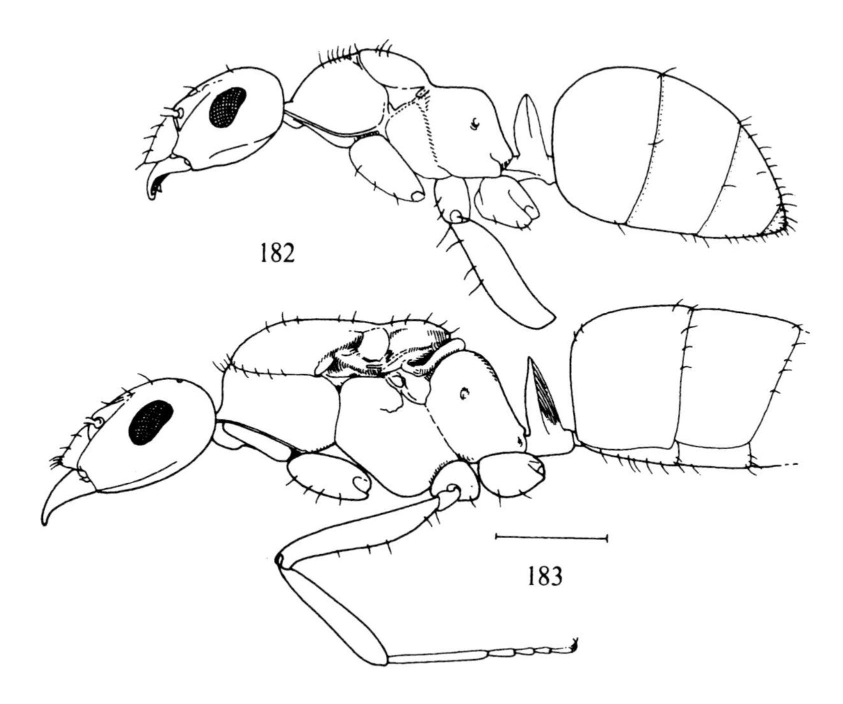

Figs. 182, 183. *Formica lemani* Bondr. – 182: worker in profile; 183: queen in profile. Scale: 1 mm.

Distribution. Throughout Fennoscandia except extreme southern areas of Finland and Sweden, not recorded from Denmark. – Abundant in British Isles except for Southeast. – Range: mountains of Spain to Japan including Himalayas, Appenines to arctic Fennoscandia.

Biology. This is an abundant upland species in Europe not distinguished from *F. fusca* until Bondroit (1917) and first clearly described by Yarrow (1954). It has similar habits to *F. fusca* but sometimes occurs in large multi-queened colonies in favourable sites such as stony banks. Colony founding is either by single queens or by nest fission. Alatae fly in July and August.

46. ***Formica transkaucasica*** Nasonov, 1889
Figs. 163, 184–187.

Formica picea Nylander, 1846b: 1059 (jun. hom. of *Formica picea* Leach, 1825).
Formica transkaucasica Nasonov, 1889: 21; Yarrow, 1954: 232.

Worker. Shining brownish black or black. Gaster pubescence very dilute, almost absent. Frontal triangle smooth without sculpture. Pronotum with numerous long erect hairs, gula and mid femora normally with one or two long hairs. Third antennal segment only slightly longer than wide. Length: 4.5–6.0 mm.

Queen. Shining black. Sculpture, pilosity and colour as worker. Third antennal segment only slightly longer than wide. Length: 8.0–9.0 mm.

Male. Shining black; frontal triangle smooth. Adpressed pubescence on gaster very long but not obscuring cuticular shine. One or two gula hairs usually present. Long hairs on side borders of scale including sides of dorsal crest which is flat, not emarginate. Length: 7.5–8.5 mm.

Distribution. Denmark, Sweden and Finland generally distributed; Norway local, recorded from Hedmark only (Collingwood, 1976). – Very local in South England. – Range: Pyrenees to Japan, Appenines to arctic Sweden.

Biology. In mountains and northern Europe this species is restricted to sphagnum mires and wet peaty meadows. Nests are often situated in grassy tussocks with a built up cone of sphagnum and grassy fragments. In Central Asia including Mongolia and parts of the Himalayas a morphologically indistinguishable form of this species is abundant but inhabits an entirely different biotope on dry stony ground. However, according to Kutter (1977) *F. »picea«* also nests on dry land in the High Alps. Alatae occur in July.

47. ***Formica cinerea*** Mayr, 1853
Figs. 162, 188–192.

Formica cinerea Mayr, 1853: 280.

Worker: Brownish black often with genae and mesopleural articulations brownish red. Whole body closely covered with silvery pubescence. Erect hairs numerous on all dorsal surfaces, also on femora, on occiput and on gula. From above occipital hairs extend round the posterior margin of the head to the eyes. Length: 4.0–6.5 mm.

Queen. As worker. Length: 8.0–9.0 mm.

Male. Colour and pilosity as queen; legs and external genitalia yellowish to brown. Length: 7.0–8.0 mm.

Distribution: Locally abundant on coastal sand of Jutland in Denmark; Skåne, Blekinge, Halland in Sweden, also inland – Dalarna and Västmanland; in Norway only recorded from Elverum in Hedmark (Collingwood, 1963); in Finland on coasts of Ostrobottnia, Nylandia and Karelia australis, inland also in Karelia borealis and Savonia borealis. – Absent from British Isles. – Range: Pyrenees to Urals, North Italy to Central Fennoscandia.

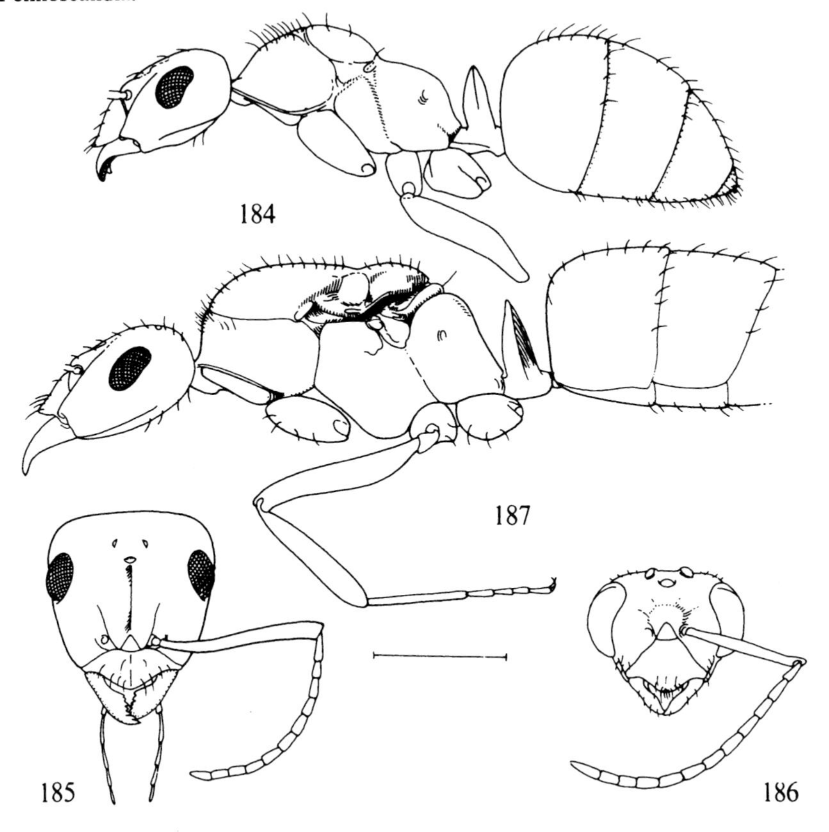

Figs. 184–187: *Formica transkaucasica* Nasonov. – 184: worker in profile; 185: head of queen in dorsal view; 186: head of male in dorsal view; 187: queen in profile. Scale: 1 mm.

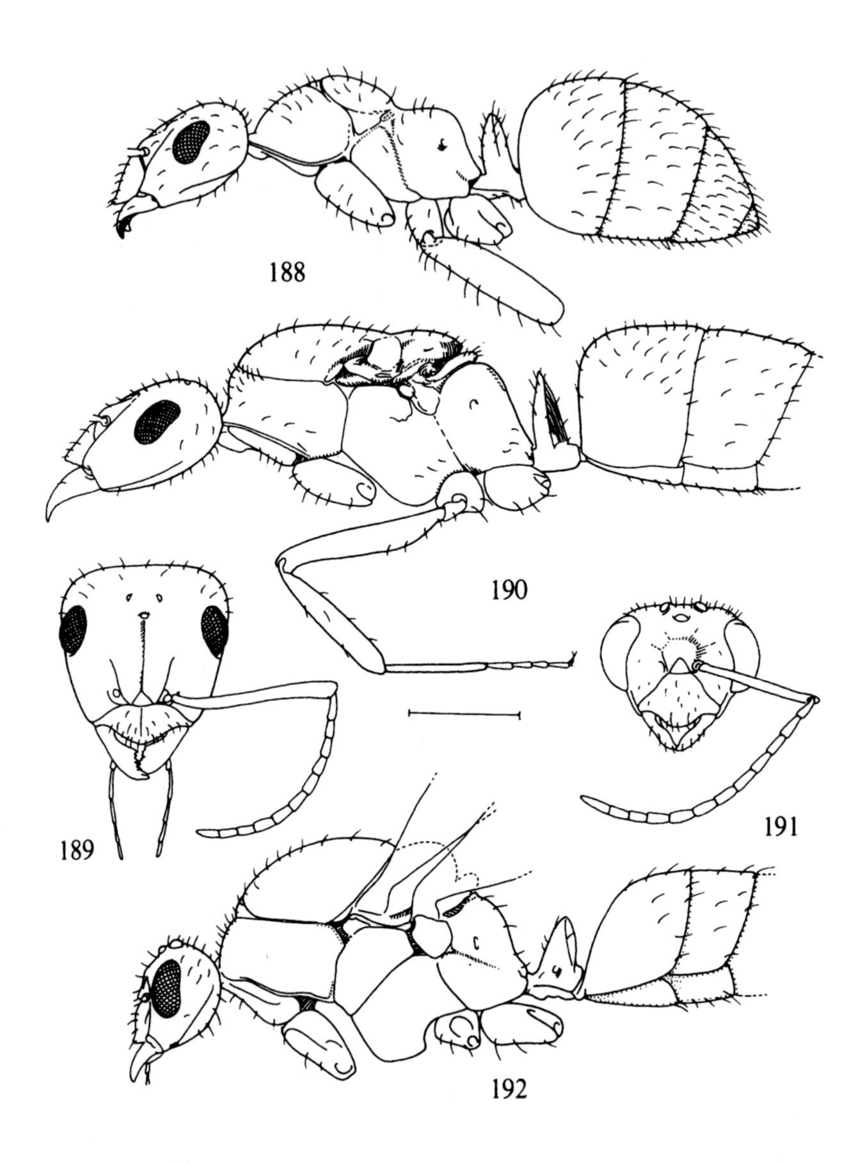

Figs. 188–192. *Formica cinerea* Mayr. – 188: worker in profile; 189: head of queen in dorsal view; 190: queen in profile; 191: head of male in dorsal view; 192: male in profile. Scale: 1 mm.

This species characteristically occurs in drift sand on coastal dunes in North Europe but also locally inland on coarse morainic drift. It is an aggressive species living largely by predation. Nests may be founded by single queens but where the species is populous, colonies are frequently polygynous and polycalic. Alatae occur in July. Its ecology and distribution in Finland are described by Kilpäinen, Valkeila, Vesajoki and Wuorenrinna (1977).

48. ***Formica cunicularia*** Latreille, 1798
Figs. 161, 167, 193–196.

Formica cunicularia Latreille, 1798: 151; Yarrow, 1954: 231 (redesription).

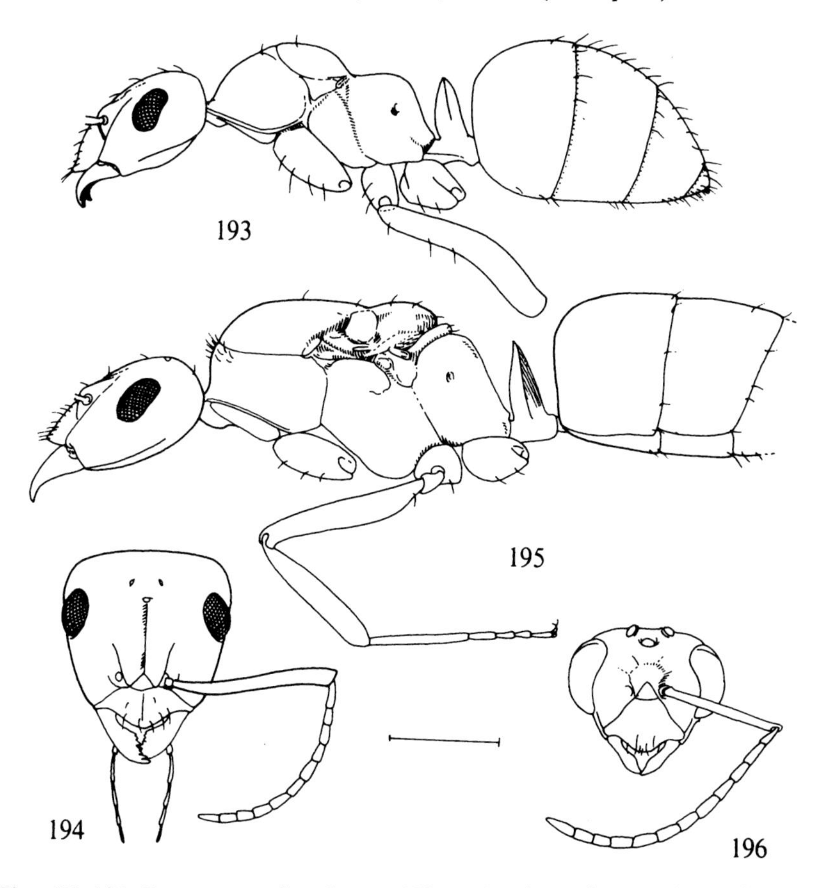

Figs. 193–196. *Formica cunicularia* Latr. – 193: worker in profile; 194: head of queen in dorsal view; 195: queen in profile; 196: head of male in dorsal view. Scale: 1 mm.

Worker. Ashy grey black with at least genae and mesopleural articulations reddish; often most of alitrunk and head may be reddish. Gula and occiput bare. Erect hairs normally absent on pronotum but occasionally one or two short erect hairs may be present on promesonotum, never on upper margin of scale. Length: 4.0–6.5 mm.

Queen. Yellowish red to dark red with most of head, mesonotum and gaster dark. Propodeum black or yellowish red. First gaster tergite often reddish in part. Erect hairs restricted to anterior part of pronotum, absent from propodeum and upper margin of scale. Appendages pale to dark brownish red. Length: 7.5–9.0 mm.

Male. Body uniformly dark; gaster and scutellum dull with close pubescence; legs mainly yellowish red. Long hairs present on dorsal margin of scale which is emarginate with pronounced angular corners. Length: 8.0–9.0 mm.

Distribution. Very local – Denmark: only East Jutland; Sweden: Skåne, Blekinge and Halland in the Southwest, also in Öland, Gotland and Gotska Sandön. – Locally common in South England and South Wales. – Range: North Africa to South Scandinavia, Portugal to Urals.

Biology. This is a common species throughout Western Europe, nesting under stones or in small earth mounds, colonising railway embankments, sun exposed borders of woodland, dry open pasture and sea cliffs. Each nest is separate and normally has only one queen. Its habits are mainly predaceous and scavenging. Alatae occur in July and August.

49. ***Formica rufibarbis*** Fabricius, 1793
Figs. 160, 168, 197–200.

Formica rufibarbis Fabricius, 1793: 355.

Worker. Head and alitrunk mainly red with variable amounts of dark on promesonotum and hind part of head. Gaster thickly pubescent, dull. Erect hairs numerous on pronotum and normally present on upper margin of scale, absent on gula and occiput. Length: 4.5–7.0 mm.

Queen. Brightly coloured with red predominating. Scutellum usually black, gaster dull. Upper margin of scale and propodeum with erect hairs. Pronotal hairs numerous, extending round side margins to tegulae. Length: 8.5–10.5 mm.

Male. Blackish brown. Scutellum and gaster pubescent and dull. Petiole crest emarginate with sharp side angles and numerous long hairs. Femora mainly dark. Length: 8.5–9.5 mm.

Distribution. Locally common throughout Denmark and Southern Fennoscandia to latitude 62°. – Very local in South England. – Range: Portugal to Western Siberia, Mountains of Middle East toSouth Fennoscandia.

Biology. This is a widely distributed species occurring throughout Europe, nesting in the ground with a single entrance hole or under stones. It is predatory and aggressive and readily attacks other species of ants and insects. New nests are started by single queens alone. Mature colonies are separate but may contain two or three queens with up to 500 or more workers. Alatae fly in late June and July.

50. ***Formica exsecta*** Nylander, 1846
Figs. 201, 202, 208, 212, 214, 217.

Formica exsecta Nylander, 1846a: 909.

Worker. Bicoloured with gaster dark brown, rest of body reddish with varying amount of dark colour on head and promesonotum. Head strongly excised posteriorly; max-

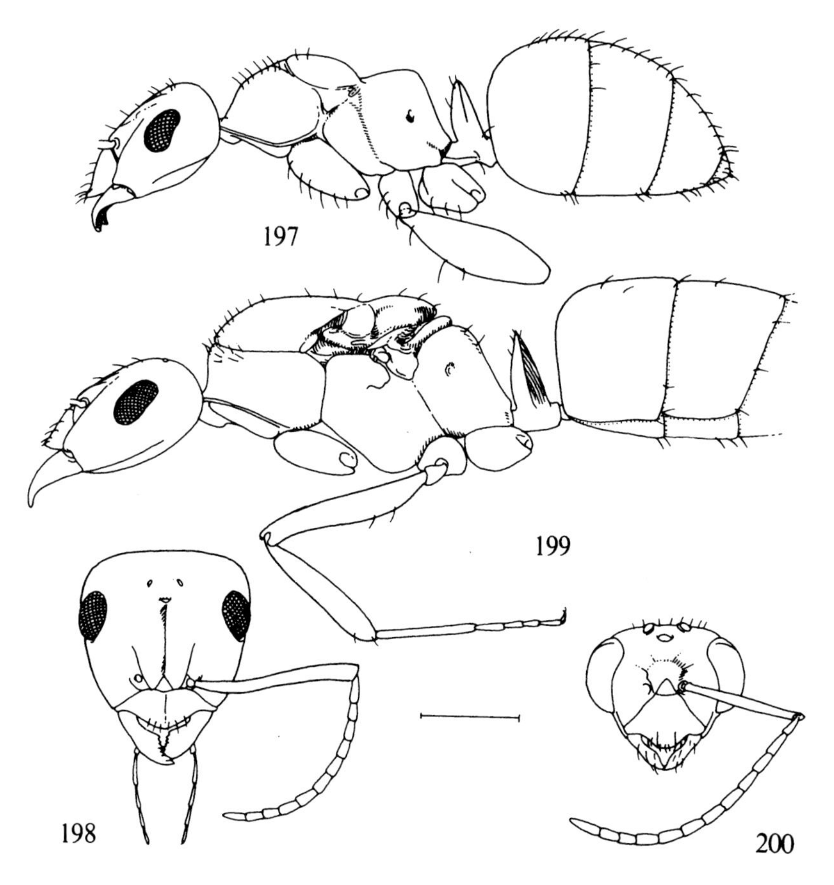

Figs. 197–200. *Formica rufibarbis* Fabr. – 197: worker in profile; 198: head of queen in dorsal view; 199: queen in profile; 200: head of male in dorsal view: Scale: 1 mm.

illary palps 6 segmented, long as half head length. Scale strongly emarginate. Eyes with very distinct erect hairs which are normally abundant. Body pilosity variable – erect hairs on all gaster tergites, on clypeus and on dorsum of head, sometimes also on occipital margins. Clypeus not impressed. Length: 4.5–7.5 mm.

Queen. As worker, head normally somewhat darker and promesonotum more or less dark brown. Head pilosity very variable but eyes always distinctly haired. Length: 7.5–9.5 mm.

Male. Dark brownish black, appendages yellowish to brown. Head broadly emarginate, scale excised. Eyes with distinct but sparse hairs. Maxillary palps 6 segmented, long. Length: 6.2–9.0 mm.

Distribution. Throughout Denmark and Fennoscandia, very common. – Local in Southwest England and Scottish Highlands. – Range: Central Spain to Urals, Appenines to extreme north of Europe.

Biology. This is an active aggressive species building mounds of leaf litter in open woodland, moorland and rough pasture. On disturbance the ants swarm out and bite vigorously. Nests may contain a thousand or more workers with more than one queen. They are often grouped with amicable interchange of workers between each. *F. exsecta* is mainly aphidicolous tending aphids on Juniperus, Picea and other trees but is also predaceous. Colonies extend by nest splitting but single queens also start colonies by securing acceptance in nests of *Formica lemani* or *F. fusca*. Alatae occur in July.

51. ***Formica foreli*** Emery, 1909
Fig. 216.

Formica foreli Emery, 1909: 192.

Worker: Bicoloured with head and gaster mainly dark and with a pronounced dark patch on promesonotum. Eyes without hairs and dorsal body hairs restricted to front of clypeus and apical tergites of gaster. Maxillary palps very short either 5 or 6 segmented, not reaching back beyond front eye margin. Pubescent hairs on gaster longer than their interspace. First gaster tergite somewhat dull, not shining. Clypeus normally slightly impressed below midline when seen in profile. Head and scale strongly excised. Length: 4.5–7.0 mm.

Queen. Bicoloured with part of head, mesopleurae at sides and scale reddish yellow. Gaster dull not shining, rest as worker. Length: 6.0–7.0 mm.

Male. Black with yellowish appendages; head broadly emarginate; eyes without hairs. Mesonotum with numerous short erect hairs. Pubescent hairs on gaster slightly longer than interspaces. General appearance somewhat shining but less so than *F. pressilabris*. Length: 5.5–7.0 mm.

Distribution. So far only known from Tisvilde Hegn and Asserbo, NEZ, Denmark, but because only queens are easily distinguished from *F. pressilabris* may be underrecorded. – Range: Central Europe, rather local – France to Caucasus, Switzerland to Poland and Denmark.

Biology. This species nests in small flat mounds of grass and heather litter on banks in open lowland heath. Habits are similar to those of *F. exsecta.*

52. ***Formica forsslundi*** Lohmander, 1949
Figs. 204, 205, 210.

Formica forsslundi Lohmander, 1949: 163.

Worker. Bicoloured, appearance shining, pubescence very sparse. Eyes bare. Erect hairs are present sparsely on clypeus, head, alitrunk and on gaster tergites 2 to apex. Maxillary palps very short – 5 segmented. Head and scale deeply excised. Length: 4.0–6.5 mm.

Queen: Brownish, brilliantly shining. Head and scale deeply excised. Rest as worker. Size very small relative to worker. Length: 5.0–6.0 mm.

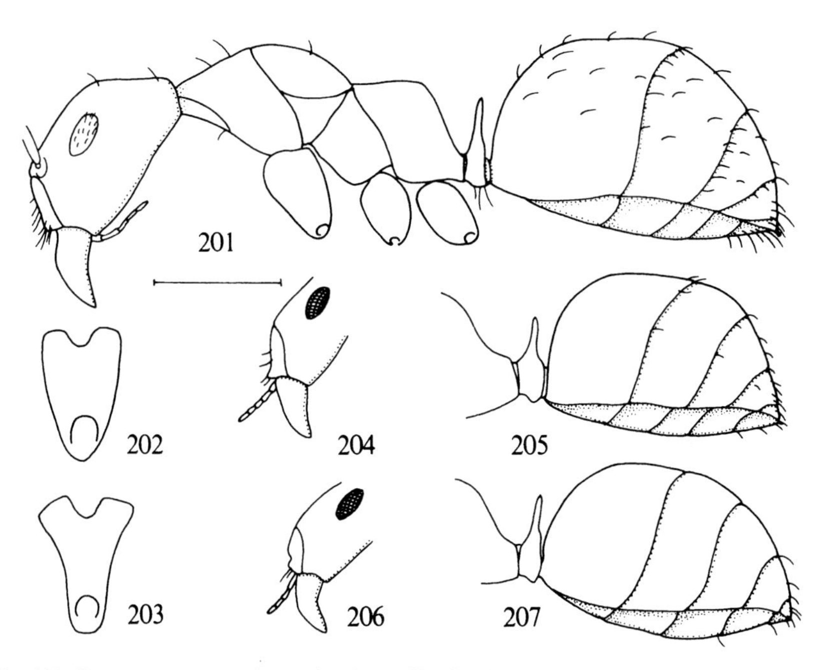

Fig. 201. *Formica exsecta* Nyl., worker in profile. Scale: 1 mm.
Figs. 202, 203. Petiole scale of *Formica*-queens in anterior view. – 202: *exsecta* Nyl.; 203: *suecica* Adlerz. Scale: 1 mm.
Figs. 204, 205. *Formica forsslundi* Lohm. – 204: head of worker in profile; 205: gaster of worker in profile. Scale: 1 mm.
Figs. 206, 207. *Formica pressilabris* Nyl. – 206: head of worker in profile; 207: gaster of worker in profile. Scale: 1 mm.

Male. Black. Mesonotum and gaster tergites with long hairs. Eyes bare. Head and mesonotum finely sculptured, scale and gaster distinctly shining. Maxillary palps very short. Head and scale broadly emarginate. Length: 5.5–6.5 mm.

Distribution. Locally abundant on wet heathland and in forest mires. Denmark: EJ and WJ. – Sweden: Sm. to Hrj. – Norway: HE only. – Finland: Ab, St, Tb and Ob (north to Rovaniemi). – Range: Switzerland (very local), Denmark to Urals, very local outside Fennoscandia.

Biology. This interesting species is found exclusively in wet heathland and mires in open forest. Colonies are started by queens securing adoption with *Formica transkaucasica*. Nests are constructed of leaf litter usually on the margins of wet ditches or fringes of boggy lakes. Alatae have been found in the nests during July.

53. ***Formica pressilabris*** Nylander, 1846
Figs. 206, 207, 211, 215.

Formica pressilabris Nylander, 1846a: 911.

Worker. Bicoloured; head and scale deeply excised. Eyes bare. Maxillary palps very short, 5 or 6 segmented. Erect hairs on dorsum restricted to anterior margin of clypeus and gaster tergites 4 to apex. Clypeus transversely impressed below mid line, with a distinct concavity when seen i profile. Gaster pubescence sparse with hairs slightly shorter than their interspace – general appearance moderately shining. Length: 4.2–6.0 mm.

Queen. More or less evenly dark brown with propodeum and under body lighter, very shining, rest as worker. Size small. Length: 4.5–5.5 mm.

Male. Dark brown, somewhat shining. Mesonotum clothed in short fine hairs. Erect hairs on gaster restricted to apical segments. Pubescent hairs sparse about as long as interspace. Length: 4.5–6.0 mm.

Distribution. Local in South Norway. Locally abundant in Denmark, Sweden and Finland as far north as Med. and Ks, respectively. – Not found in British Isles. – Range: widely distributed from Spanish Pyrenees to Siberia, Italy to Central Fennoscandia.

Biology. This species constructs football size mounds of grass litter in dry pasture and on banks in open woodland. Usually two or more nests are found together with up to two thousand workers and several queens in each. In Poland *F. pressilabris* has been extensively studied by Czechowski (1975); there the species is mainly found in open meadows in polygynous polycalic colonies of many nests. The chief food source was the exudate of species of aphids feeding on herbage and very little predatory activity was observed. In Fennoscandia nests observed have usually been either single or more commonly in groups of up to five. Although similar in appearance to *F. forsslundi* this is a dry habitat species and does not normally occur in the neighbourhood of mires. Alatae occur in July and August.

54. ***Formica suecica*** Adlerz, 1902
Figs. 203, 209, 213, 218.

Formica suecica Adlerz, 1902: 263.

Worker. Alitrunk and head reddish, ocellar region often indistinctly brownish; gaster brown with basal face reddish. Head broad with rounded sides and occipital corners which round gently into shallow posterior emargination. Scale with rounded dorso-lateral angles and flat central emargination. Palpi six segmented, short not extending beyond front eye margin. Scattered hairs on clypeus, frons and dorsum of all gaster tergites. Eyes without hairs. Length: 4.5–6.5 mm.

Queen. Brownish black with only propodeum paler reddish brown. Head broad with sides and occipital angles broadly rounded. Scale with distinct dorso-lateral lobes. Eyes bare. Length: 5.5–6.3 mm.

Male. Dark brown, legs and genitalia pale brown. Head broadly rounded, scarcely emarginate posteriorly; dorsal surfaces with scattered erect hairs; eyes bare. Maxillary palps moderately long, 6 segmented. Length: 6.0–7.0 mm.

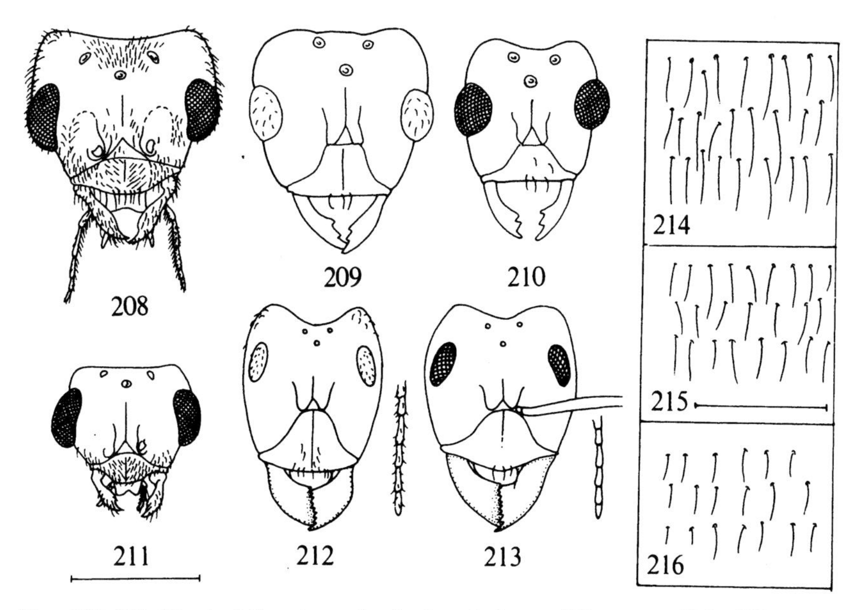

Figs. 208–211. Head of *Formica*-males in dorsal view. – 208: *exsecta* Nyl.; 209: *suecica* Adlerz; 210: *forsslundi* Lohm:; 211: *pressilabris* Nyl. – Scale: 1 mm.
Figs. 212, 213. Head in dorsal view and maxillary palp of *Formica*-workers. – 212: *exsecta* Nyl.; 213: *suecica* Adlerz. Scale: 1 mm.
Figs. 214–216. Gaster pubescence in *Formica*-workers. – 214: *exsecta* Nyl.; 215: *pressilabris* Nyl.; 216: *foreli* Emery. Scale: 0.02 mm.

Distribution. Not in Denmark. – Sweden: generally distributed from Sm. north to Lu. Lpm., but not recorded from Öl., Gtl. and G. Sand. – Norway: HE, VE, R and SF. – Finland: Ab, N, Om, ObS and ObN. – Range: Fennoscandia and Esthonia only.

Biology. This is a truly endemic Fennoscandian species, not recorded east of longitude 30° or south of latitude 56°. Nests are in open sites in tree stumps with scattered leaf litter but not piled up into a dome. This species may be confused with the redder examples of *F. exsecta* but is easily distinguished by the broadly rounded head and bare eyes. Males and queens occur in July and the small queens start fresh colonies by adoption by either *F. fusca* or less commonly *F. transkaucasica*. The habits of *F. suecica* have been studied by Adlerz (1902) on the offshore island of Ålnö in Central Sweden where many nests were found, by Holgersen (1943) in Norway and by Forsslund (1947).

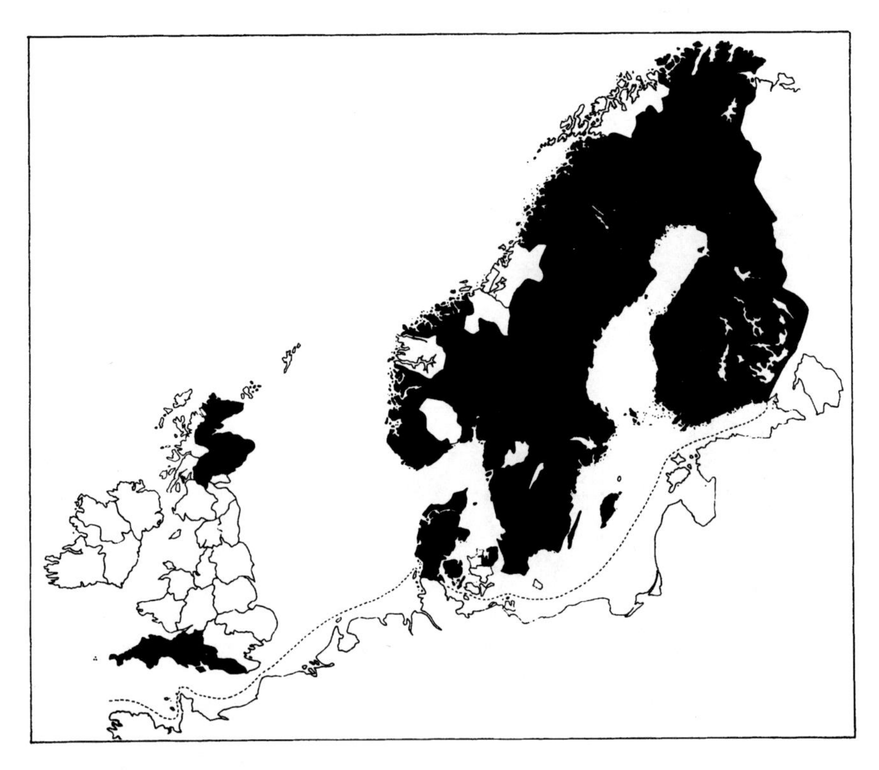

Fig. 217. Distribution of *Formica exsecta* Nyl., a North and Central European species.

55. ***Formica uralensis*** Ruzsky, 1895
Figs. 219, 220.

Formica uralensis Ruzsky, 1895: 13; 1896: 69 (German translation).

Worker. Head entirely black, dark area on dorsum of promesonotum dense black, gaster black, rest of alitrunk and appendages yellowish to brownish red. Head as broad as long, antennal scape broad and short. Frontal triangle sculptured and dull. Bristlelike hairs on dorsum of head, gula, alitrunk and gaster usually present but variable in number. Length: 4.5–8.0 mm.

Queen. As worker but with whole of mesoscutum dark. Legs pitchy. Frontal triangle sculptured and dull; eyes bare. Length: 9.0–11.0 mm.

Male. Head, mandibles, antennal scapes, alitrunk and gaster dense black. Mandibles denticulate with up to 5 teeth but variable. Clypeus, head, promesonotum and scale with widely spaced hairs; eyes bare. Wings dusky, frontal triangle dull. Length: 9.0–11.0 mm.

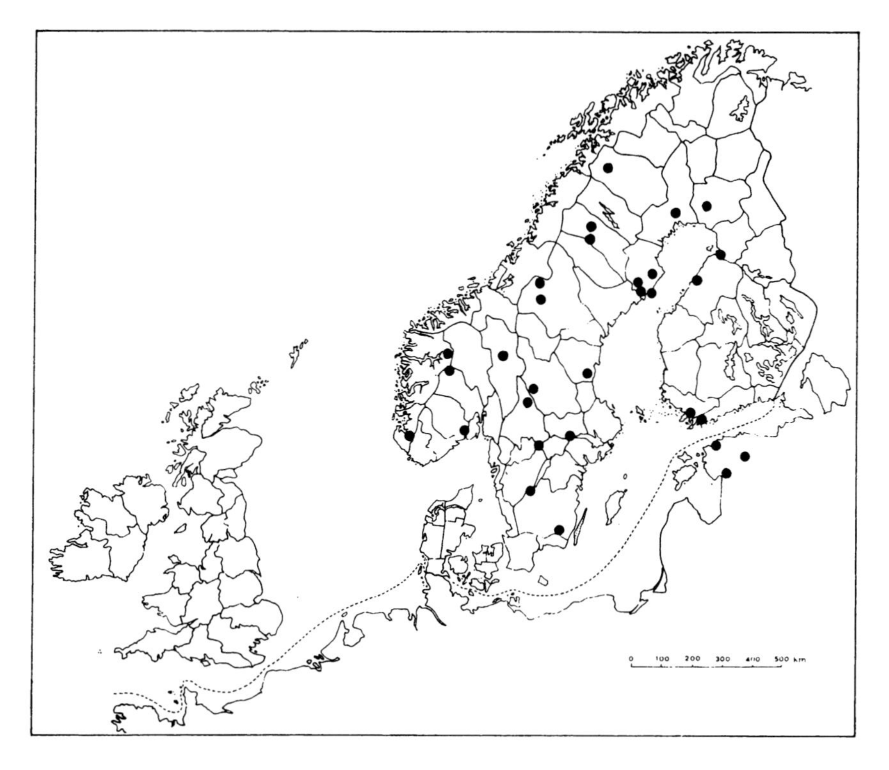

Fig. 218. Distribution of *Formica suecica* Adlerz, an endemic species.

Distribution. Local in Denmark: SJ, EJ, WJ, NEJ, and Norway: HE and Fø (Fjellberg, 1975). – Widely distributed in Sweden and Finland. – Range: Northeast Europe including N. Germany, Baltic States and West USSR; one record from Swiss Alps. Widely distributed in Mongolia and Central Siberia.

Biology. In Europe *F. uralensis* is typically found on lowland open mosses with scattered trees, more occasionally on drier heath. Nests may be isolated or in groups and are built up of leaf litter and twigs into rounded dome. The nest surface is of fine material which covers a large brood incubation chamber resting on a surface of coarse long twigs. Rosengren (1969) has studied its habits in South Finland; unlike members of the *F. rufa* group, this species does not go deep within the nest to hibernate but the ants clump together under peat moss or among tree roots away from the summer nest. Food is mainly honey dew from surrounding betula scrub or pines. Although this species has superficial similarities to *F. rufa* group species it is morphologically well differentiated with its broad black head, short thick antennae and wide coarsely sculptured frontal triangle.

Nests are usually polygynous and may reproduce by colony fission but fresh colonies may also originate from adoption of fertile queens by *F. transkaucasica.* Alatae occur in July. Its marshy habitat in Europe contrasts with the dry steppe habitat in Asia and may be related to the inability of this species to survive aggressive competition from other wood ant species since according to Rosengren (1969), although *F. uralensis* defends its terrirory it is easily overwhelmed by other ants such as *F. sanguinea* and *Myrmica rubra.*

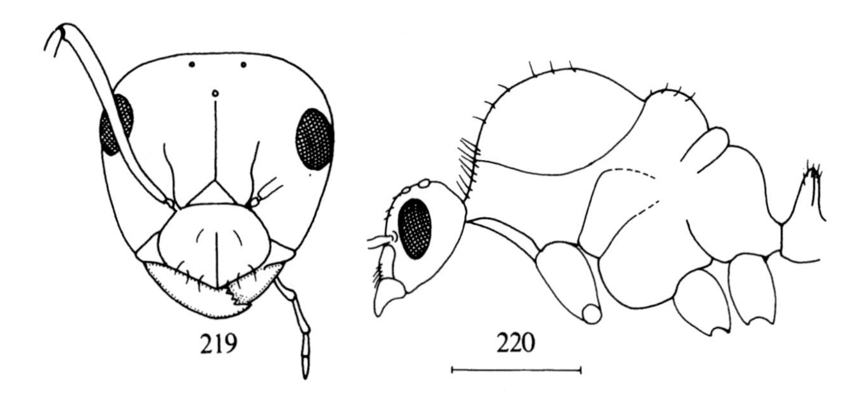

Figs. 219, 200. *Formica uralensis* Ruzsky. – 219: head of worker in dorsal view; 220: head and alitrunk of male in profile. – Scale: 1 mm.

56. ***Formica sanguinea*** Latreille, 1798
Figs. 221–223.

Formica sanguinea Latreille, 1798: 37.

Worker. Alitrunk and front of head bright red; gaster and ocellar region of head brownish black. Proportion of dark to red varies from north to south with some arctic samples having proponderantly dark heads and dusky red alitrunk contrasting with southern samples which may have whole body excluding gaster bright red. Frontal triangle dull; eyes without hairs. Clypeus with median anterior notch variable in size and shape but always present. Head and mandibles broad relative to alitrunk and gaster. Length: 6.0–9.0 mm.

Queen. As worker with dark areas on alitrunk absent or restricted to sides of mesopleurae, but some northern samples darker. Gaster small in relation to head. Length: 9.0–11.0 mm.

Male. Black, legs yellow. Clypeal notch less distinct than in female castes. Mandibles each with four or five small teeth. Frontal triangle dull; eyes bare. Length: 7.0–10.0 mm.

Distribution. Common throughout Denmark and Fennoscandia. – Locally common in England and Scotland. – Range: throughout Eurasia from Portugal to Japan and Iran to Arctic Norway.

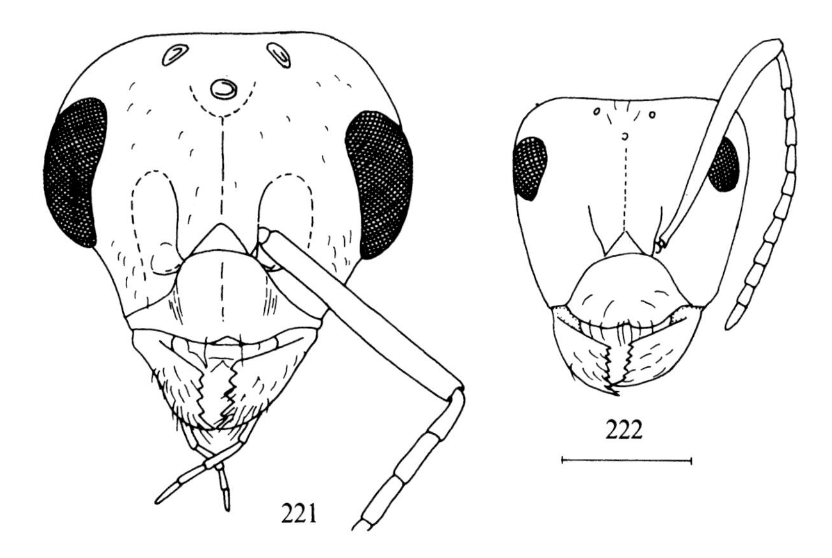

Figs. 221, 222. *Formica sanguinea* Latr. – 221: head of male in dorsal view; 222: head of worker in dorsal view. Scale: 1 mm.

Biology. This is the well known aggressive red slave-making ant raiding nests of any species in its neighbourhood during the summer and removing pupae of the *Formica fusca* group of species for rearing in the parent nest as auxiliaries and also as food. All Fennoscandian species of the *F. fusca* group have been found in mixed colonies with *F. sanguinea* including *F. rufibarbis*, *F. cinerea*, *F. gagatoides* and *F. transkaucasica* as well as the more frequent *F. fusca* or *F. lemani*. *F. sanguinea* tends to avoid in-fighting but overwhelms other species by abrupt aggressive movements. Nests are situated under stones or in tree stumps with a small accumulation of leaf litter. It is often a dominant species in cleared woodland and in some localities all other *Formica* species have been eliminated and in such cases only pure *F. sanguinea* colonies are to be found, usually with a high proportion of small workers to act as nurses. Colonies spread by nest splitting and also by individual queens entering nests of the auxiliary species and appropriating a part of the brood, the host queen or queens being subsequently destroyed. Alatae are developed in July pairing often occurring in the vicinityof the nest.

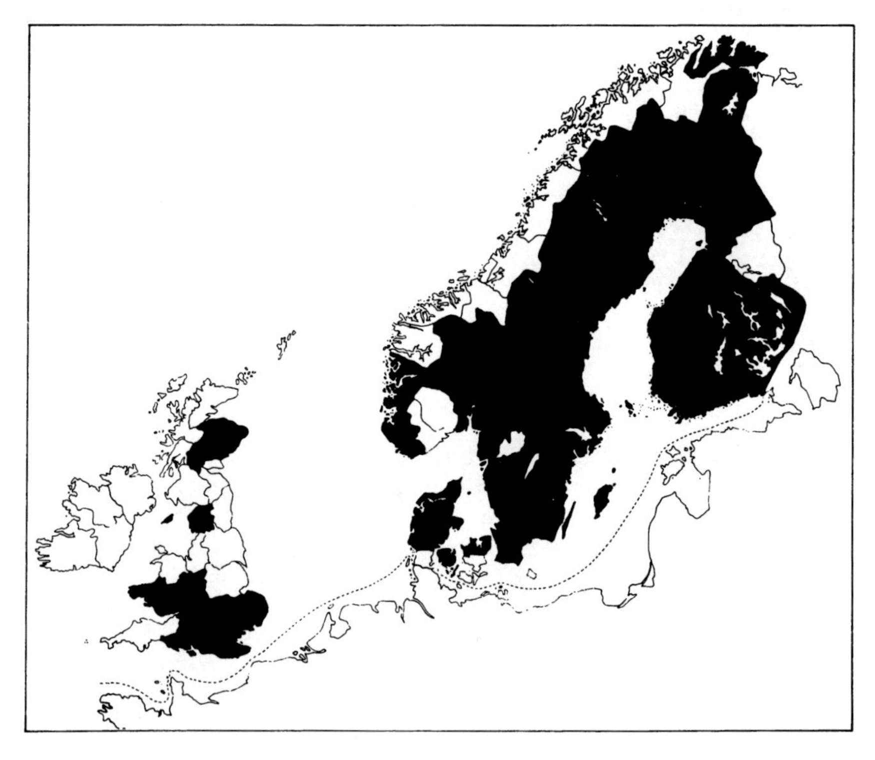

Fig. 223. Distribution of *Formica sanguinea* Latr., a wide ranging boreal species.

57. ***Formica truncorum*** Fabricius, 1804
Figs. 224–228.

Formica truncorum Fabricius, 1804: 403.

Worker. Large workers with head, alitrunk and base of first gaster tergite bright yellowish red, gaster greyish brown covered with long pubescence; smaller workers are usually darker but never with clearly marked black patches as in *F. pratensis*. Eyes, occiput, genae, gula, scapes and tibiae as well as whole body covered in short erect hairs. Frons with large shallow punctures; frontal triangle shining without punctures or sculpture. Funiculus in larger workers slender with segments two and three twice as long as wide. Lateral clypeal pits deep and rounded. Length: 3.5–9.0 mm.

Queen. Brightly coloured with head and most of alitrunk red or with part of ocellar region and most of scutum brownish. Pilosity and other features as in worker. Length: 8.0–9.5 mm.

Male. Black with appendages and external genitalia entirely yellowish. All surfaces covered with short erect hairs. Punctures coarse and shallow on head and alitrunk; frontal triangle shining without sculpture. Mandibles with three or four teeth. Length: 7.0–9.0 mm.

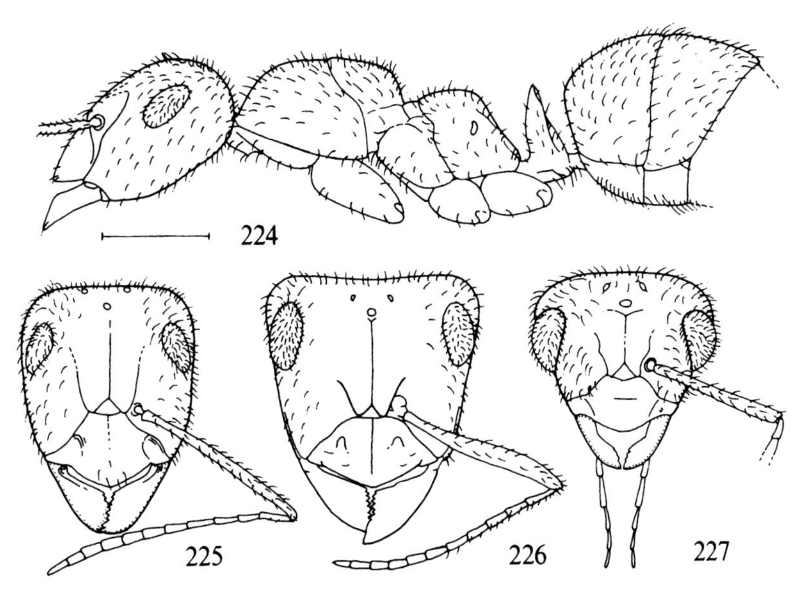

Figs. 224–227. *Formica truncorum* Fabr. – 224: worker in profile; 225: head of worker in dorsal view; 226: head of queen in dorsal view; 227: head of male in dorsal view. Scale: 1 mm.

Distribution. Locally common throughout Denmark and Fennoscandia, not found in British Isles. – Range: Jura Alps to North Japan, Italy to North Norway.

Biology. This species has large spreading colonies among stones or in tree stumps with loose surface leaf litter sometimes built into a shallow loose mound. This is an aggressive acid squirting species found at the borders of woodland and in stony banks and often particularly abundant on offshore islands. *F. truncorum* is normally polygynous, sometimes with many small dark headed queens. New colonies may be formed by nest splitting or by the adoption of single large red headed queens by *F. fusca* and allied species. Males and queens occur in July and August, latter than with most members of the *F. rufa* group.

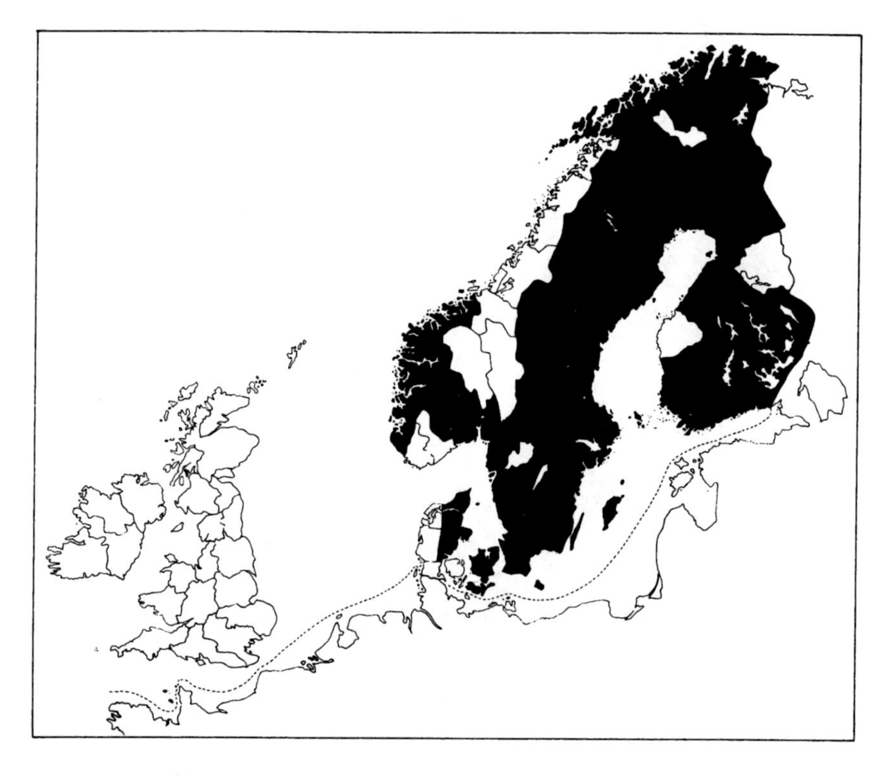

Fig. 228. Distribution of *Formica truncorum* Fabr., a Central and N. E. European species.

Formica rufa group

This is taxonomically a very difficult group with variations in sculpture and placement and abundance of body and appendage hairs. All species are both predaceous and aphidicolous. This group of ants has been the subject of a very large number of papers many dealing with the supposed significance of particular species in maintaining forest hygiene by destroying arthropod pests of trees, but both beneficial predators as well as harmful insects are taken and the ants themselves are seldom evenly distributed throughout a forest area (Adlung, 1966). This species group is also of particular interest as hosts to a number of myrmecophilous insects, arthropods and crustaceans many of which may be local or rare and entirely restricted to the wood ant nest habitat.

58. *Formica rufa* Linné, 1761

Figs. 169, 171, 229–235.

Formica rufa Linné, 1761: 426; Yarrow, 1954 (redescription).

Worker. Bicoloured red and brownish black with variable degree of depth and size of dorsal dark patch on head and promesonotum. Eyes usually with a few microscopic hairs. Long erect hairs more or less abundant on gula, clypeus, dorsum of head, alitrunk, scale and gaster but never on scape nor on posterior border of occiput. Occasional to few suberect hairs on extensor surfaces of hind tibiae and femora. Frontal triangle reflecting light but often in part with micropunctures. Frons somewhat shining with widely spaced indistinct fine punctures and scattered coarse punctures; coarse and fine punctures widely spaced on disc of first gaster tergite. Funiculus segments two and three always less than twice as long as wide. Length: 4.5–9.0 mm.

Queen. Bicoloured with whole of scutum, gaster and part of occipital area of head normally dark. Scutellum and gaster distinctly shining, never dull. No erect hairs on posterior border of occiput and normally entirely absent on basal face of gaster. Pubescence on gaster short and sparse. Micropunctures interspersed with larger punctures on disc of first gaster tergite always widely spaced. Length: 9.5–11.0 mm.

Male. Black with most of appendages and genitalia paler. Eyes with few or many erect hairs. Genae below eyes normally without outstanding hairs. Dorsum of head and alitrunk with suberect hairs. Sparse suberect hairs visible at dorsolateral margins of all gaster tergites. Upper margin of hind femora without row of short hairs. Frontal triangle shining with or without micropunctures. Length: 9.0–11.0 mm.

Distribution. Generally common throughout Denmark and Fennoscandia north to approximately latitude 63°. – Locally common in South England and Wales, more local in North England.

Biology. This is the common wood ant of most of lowland Europe building large hill nests of leaves and twigs. Nests may be isolated or in small groups, normally with many queens, up to 100 or more. Various estimates of numbers of workers in a populous nest range from 100,000 to 400,000. This is an aggressive acid squirting but somewhat

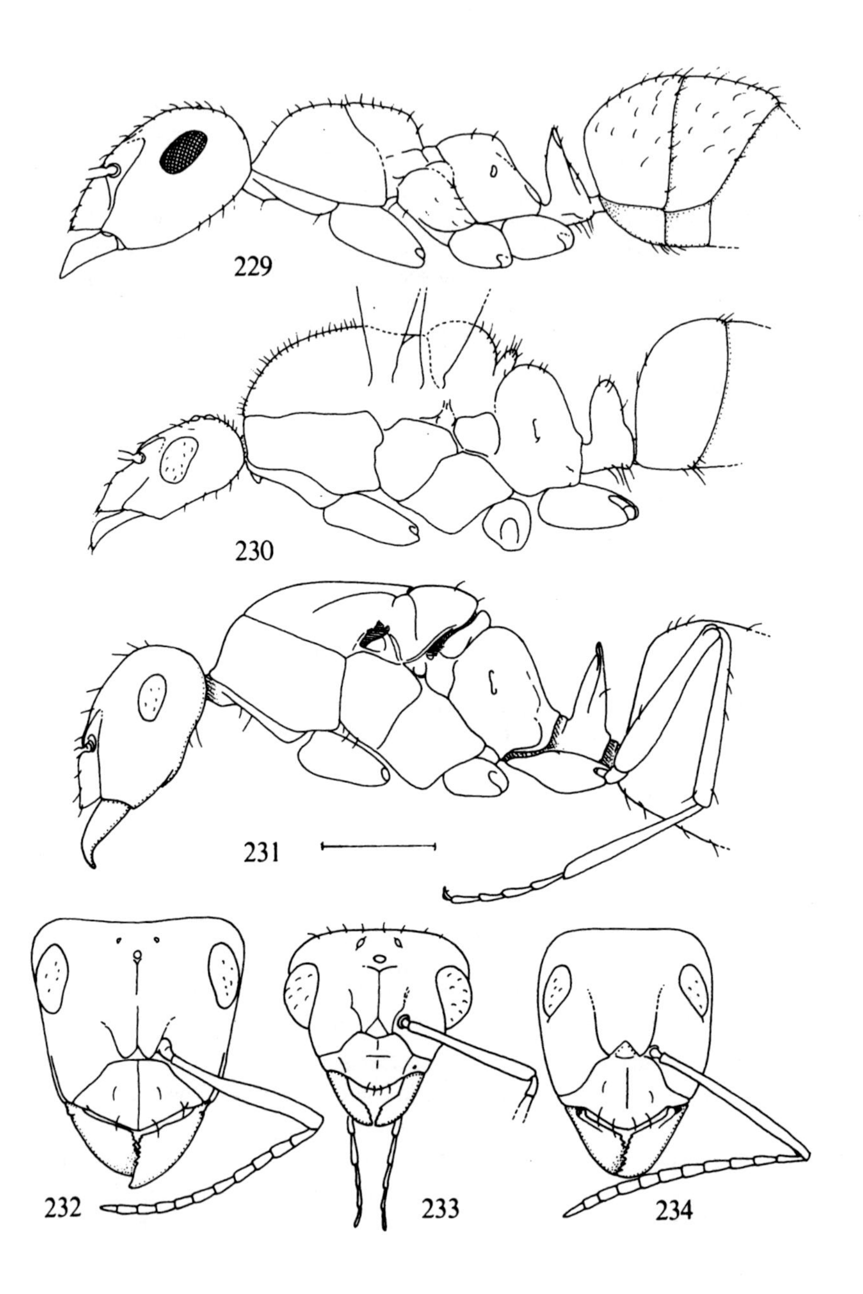
229
230
231
232
233
234

clumsy species. Foraging trails radiating from a large nest may be up to 100 m or more usually orientated toward suitable aphid bearing trees. Prey is taken somewhat unselectively from both trees and forest floor with any insect, arthropod or earthworm taken back as food to the nest although the main diet as with all species of this group is aphid honey dew. These ants mass in tight clusters on the top of the mound in the early spring sunshine. The first queen laid eggs develop into alate sexuals which fly off the nest early in the summer from May until early July. New nests arise from colony splitting in the spring but occasionally single queens may secure adoption in nests of *Formica fusca*.

A monogynous form of *F. rufa* occurs in continental Europe and probably locally in Sweden but has never been found in England. This is the *Formica rufa rufa* of Gösswald (1941). Average worker size of this form is generally large and samples are usually conspicuously hairy. Many males from such isolated colonies in the Netherlands may have one or more coarse hairs protruding from the genae below the eyes and very occasionally queens may have very short sparse hairs on the basal face of the gaster but all such individuals have the widely spaced puncturation of true *F. rufa* and although workers may have a few hairs on the upper surface of the hind femora they never form a close fringe as in other species. This form has also sometimes been referred to as *F. piniphila* Schenck, 1852. Although generally monogynous, similar hairy specimens occur on the coastal dunes of the Netherlands in polygynous colonies.

Fig. 235. Distribution of *Formica rufa* L., a southern boreal species.

Figs. 229–234. *Formica rufa* L. - 229: worker in profile; 230: male in profile; 231: queen in profile; 232: head of queen in dorsal view; 233: head of male in dorsal view; 234: head of worker in dorsal view: Scale: 1 mm.

59. ***Formica polyctena*** Förster, 1850
Figs. 172, 236–241.

Formica polyctena Förster, 1850: 15.

Worker. Erect hairs on head and alitrunk very sparse and short or absent, except on posterior margins of mesopleura. Gula hairs, if present, are restricted to one or two very weak hairs. Microsculpture is usually slightly coarser than in *F. rufa* but punctures and micropunctures are widely spaced as in that species. Length: 4.0–8.5 mm.

Queen. As *F. rufa* but middle of scutellum less shining, with fine longitudinal striae and punctures. Frons distinctly shining but rest of body due to coarser sculpture presenting a somewhat more matt appearance than in *F. rufa*. Length: 9.0–10.5 mm.

Male. Eye hairs very sparse and short. Erect hairs on promesonotum sparse, very short or absent on metanotum. Punctures on gaster and frons shallow and widely spaced. Length: 9.0–10.5 mm.

Distribution. Local, Denmark: EJ, NEJ, NEZ. – Sweden: Sk., Bl., Hall., Gtl., Nrk., Sdm., Upl., Vrm. – Norway: Ø, AK, VE. – Finland: N, Ka, Ta. – Not found in England. – Range: Spain to Siberia, Italian Alps to latitude 60° in Sweden.

Biology. This is accepted as a good species by most European authors, eg. Betrem (1960), Dlussky (1967), Kutter (1977). Some samples of *F. rufa* tend to approach the hairless condition of *F. polyctena* however, making certain determination sometimes difficult. Elton (priv. communication) found that *F. polyctena* in its most typical form readily accepted fertile queens and pupae from other distant nests of the same species but were always antagonistic to and rejected such from both polygonous and monogynous colonies of *F. rufa*. This is usually found in a group of nests and always has many queens, sometimes up to 1,000 or more.

60. ***Formica aquilonia*** Yarrow, 1955
Figs. 174, 242–248.

Formica aquilonia Yarrow, 1955a: 29.

Worker. Bicoloured with dark markings on head and promesonotum varying in size and intensity – generally not as brightly coloured or as large as *F. rufa*. In the typical form distinct outstanding hairs fringe the posterior border of the head but do not occur forward towards the eyes as in *F. lugubris*. In many samples from South Norway and South Finland these hairs may be hard to find or absent. Erect hairs on gula and dorsum of alitrunk variable, usually short and sparse. Eyes with distinct short hairs but much less prominent than in *F. lugubris*. Frons rather dull with close dense microsculpture, gaster very closely punctured. Suberect hairs on extensor surfaces of hind femora and tibiae always present but sometimes few. Antennal scapes bare. Head width of largest workers less than 2 mm. Length: 4.0–8.5 mm.

Queen. Bicoloured, scutellum rather dull but gaster always shining. In normal samples short hairs project from the posterior border of the head and on the basal face of

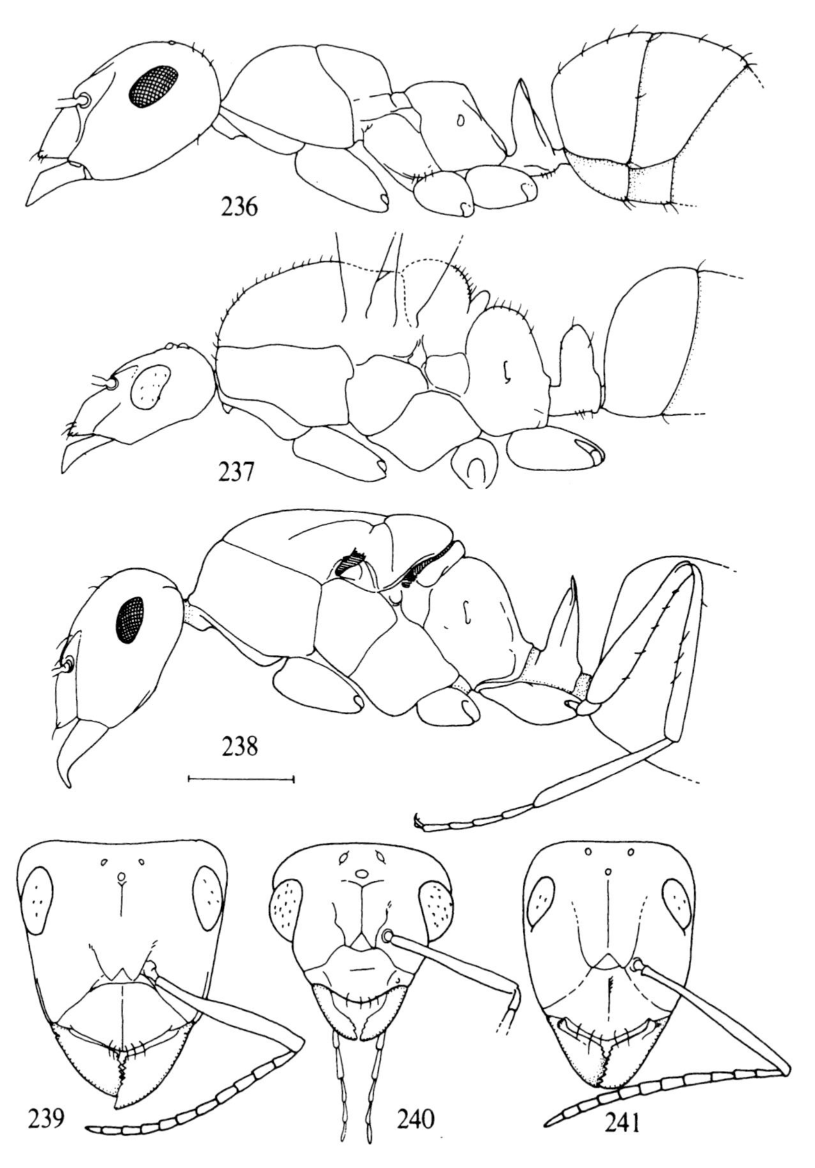

Figs. 236–241. *Formica polyctena* Förster. – 236: worker in profile; 237: male in profile; 238: queen in profile; 239: head of queen in dorsal view; 240: head of male in dorsal view; 241: head of worker in dorsal view. Scale: 1 mm.

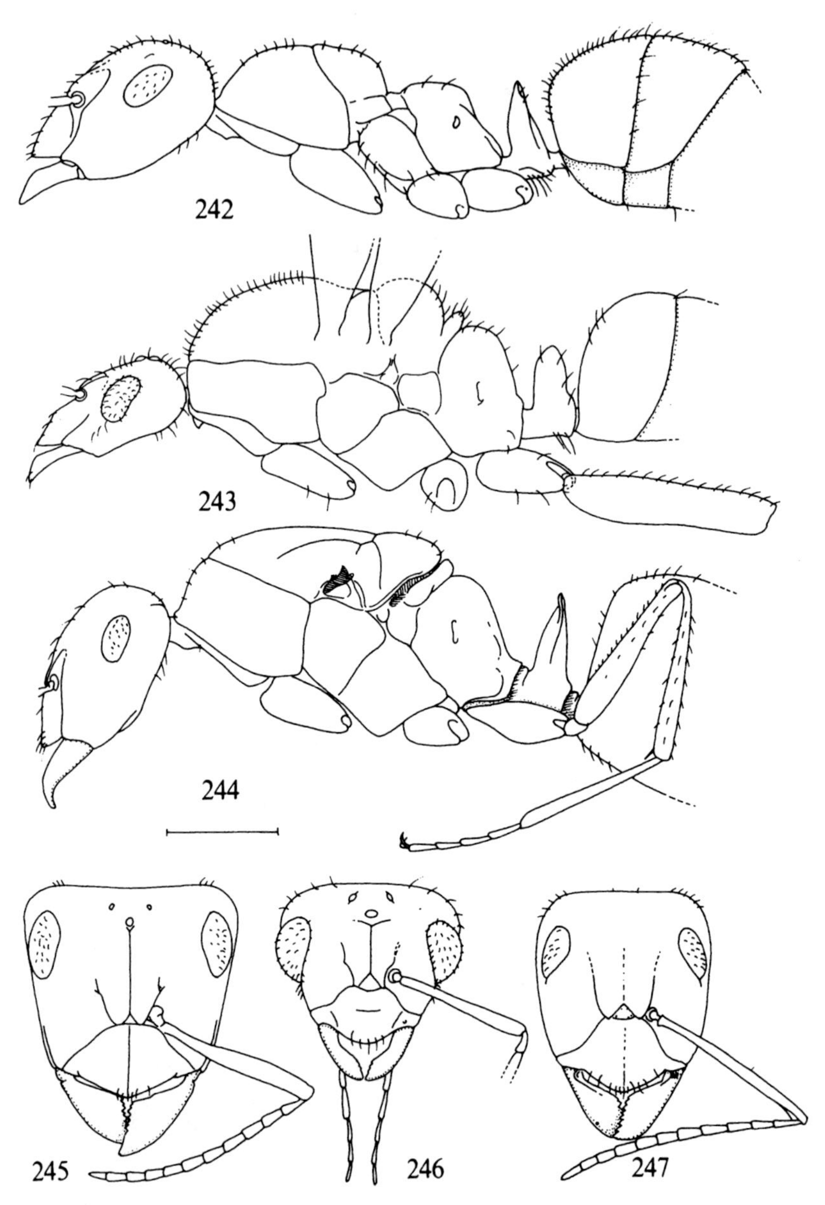

Figs. 242–247. *Formica aquilonia* Yarrow. – 242: worker in profile; 243: male in profile; 244: queen in profile; 245: head of queen in dorsal view; 246: head of male in dorsal view; 247: head of worker in dorsal view. Scale: 1 mm.

the gaster but are absent from the upper surface of the scale and propodeum. Gaster always with fine close micropunctures. Size generally smaller than other species. Length: 8.0–10.0 mm.

Male. Black, appendages paler. Genae with a few outstanding hairs below eyes. Eyes always with short hairs. Gaster tergites 2 to 4 generally without dorsolateral hairs. Fringe of short hairs always present on upper surface of hind femora. Erect hairs present on whole of head and alitrunk. Size generally smaller than other species. Length: 8.0–10.0 mm.

Distribution. Denmark: EJ, Dokkedal (Bisgaard leg.). – Sweden: Dlsl. and Vstm. northward. – Abundant throughout Norway and Finland. – Locally common in Scotland. – Range: Eastern Alps to Siberia, North Italy to North Norway.

Biology. This is undoubtedly the commonest wood ant in Fennoscandia. Large tracts of forest in the centre and north are dominated by this species which is usually found in large multicolonial groups with isolated nests being very rare. This is one of

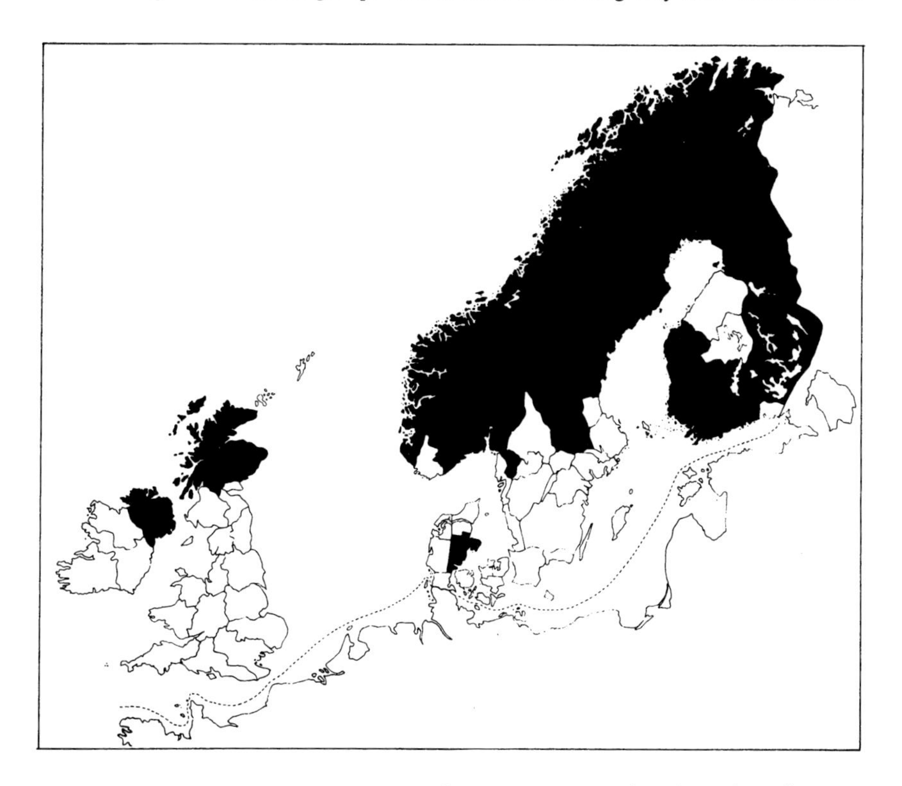

Fig. 248. Distribution of *Formica aquilonia* Yarrow, a northern boreal species.

the least aggressive species of the *F. rufa* group. Long compact trails radiate from each nest to other nests or to aphid bearing trees and antagonism between neighbouring nests has not been observed. In South Norway and South Finland comparative hairlessness in many populations makes for confusion with the rather similar *F. polyctena*. Usually, however, if enough workers are collected a majority of at least 60 % of individuals will be found to have some projecting hairs at the back of the head.

However there is a form of this species found locally in the western suburbs of Helsinki to the Sjuntio district of Ab, within an area of about 40 km by 10 km, which is almost completely hairless in all parts of the body given as species specific by Yarrow (1955) for *F. aquilonia*. All castes moreover tend to be somewhat larger and more brightly coloured. This could well be a subspecies or species in the making. Its foraging habits have been studied by Rosengren (1971, 1977a, 1977b) under the name of *F. polyctena*. This consistent degree of hairlessness has not been found elsewhere within the range of *F. aquilonia*, except perhaps in Esthonia, according to samples sent to H. Wuorenrinne by Professor V. Maavara.

The characteristics are as follows: only about 5 % or fewer workers in a series have an indication of short hairs projecting from the occipital corners of the head. Queens have no such hairs but occasional microscopic hairs have been detected on the basal face of the gaster in a very few specimens of about 50 examined. In the few males examined only 1 in 12 has projecting genal hairs. The reasons for retention as an infraspecific form of *F. aquilonia* include the close sculpturing of the worker frons, small eye hairs which are always present as in *F. aquilonia* while hairs on the extensor surface of the femora form a more or less close fringe as in *F. aquilonia* in a majority of the workers. The queen, which appears more brilliant than *F. aquilonia*, has extremely close micropunctures on the gaster as in that species (and in this character alone is quite unlike *F. polyctena*) while the male has fringing femoral hairs although specimens are also somewhat larger and more shining than *F. aquilonia*.

61. ***Formica lugubris*** Zetterstedt, 1840.
Figs. 173, 249–255.

Formica lugubris Zetterstedt, 1840: 449.

Worker. Bicoloured with distinct but not well demarcated dark patch on promesonotum. Frontal groove distinctly shining. Large punctures coarse and deep, widely dispersed among close set microscopic puncturation. Occiput with a thick fringe of hairs extending forward over area between ocelli and sides of head and laterally round to the eyes. Eye hairs erect and prominent. Body pilosity including gula, tibiae and femora more or less densely pilose. Some populations have scape hairs. Head width of largest workers 2.1 mm. Length: 4.5–9.0 mm.

Queen. Hairs and sculpture as in worker. Scale, basal face of gaster always with more or less numerous long hairs bent at the tip. Gaster, scutellum and frontal groove shining. Length: 9.5–10.5 mm.

Male. Black, legs and external genitalia yellowish to testaceous brovn. Hairs on eyes,

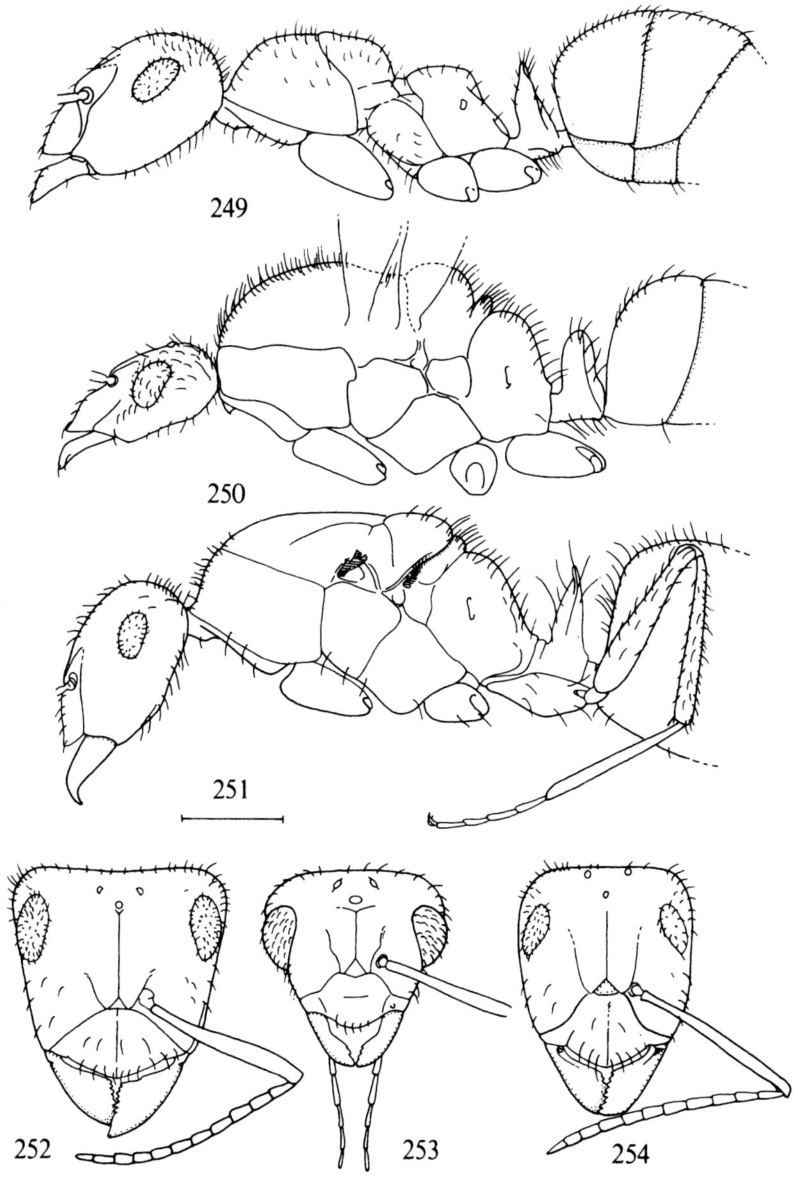

Figs. 249–254. *Formica lugubris* Zett. – 249: worker in profile; 250: male in profile; 251: queen in profile; 252: head of queen in dorsal view; 253: head of male in dorsal view; 254: head of worker in dorsal view. Scale: 1 mm.

genae below the eyes and dorsum of gaster prominent and clearly visible. Coarse punctures of head, alitrunk and gaster widely spaced among close set micropunctures. Gaster and scutellum always at least moderately shining. Length: 9.5–10.5 mm.

Distribution. Absent from Denmark and South Sweden; abundant throughout Norway and Finland and from Central Sweden northward. – Locally common in North Britain, local in South Ireland. – Range: northern Eurosiberia and European mountains from Pyrenees to Kamchatka and Japan, Italy to North Norway.

Biology. This is a robust active species. Colonies are often in groups with inter-connecting nests. It has similar habits to *F. rufa* but is able to forage at much lower temperatures and replaces *F. rufa* entirely from Central Fennoscandia to the far north. This species varies in the presence, abundance or absence of scape hairs in the female castes and some local populations in South Finland and in the Alps with such hairs have widely spaced micropunctures on the dorsum of the gaster as in *F. rufa*. Because of great variability among local populations in these areas it has not been possible to

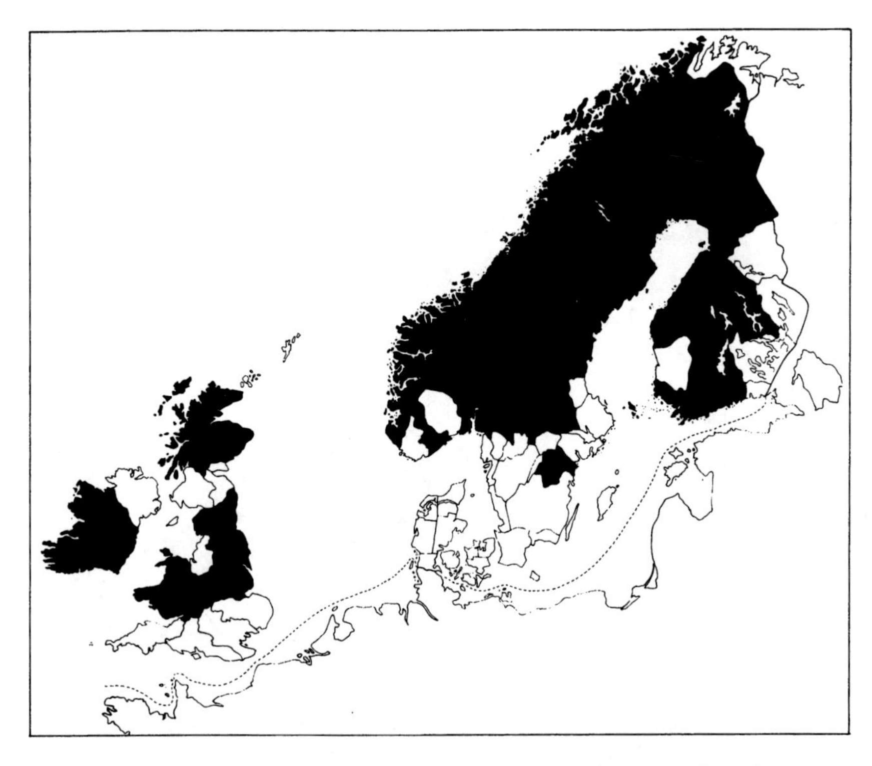

Fig. 255. Distribution of *Formica lugubris* Zett., a northern boreal species.

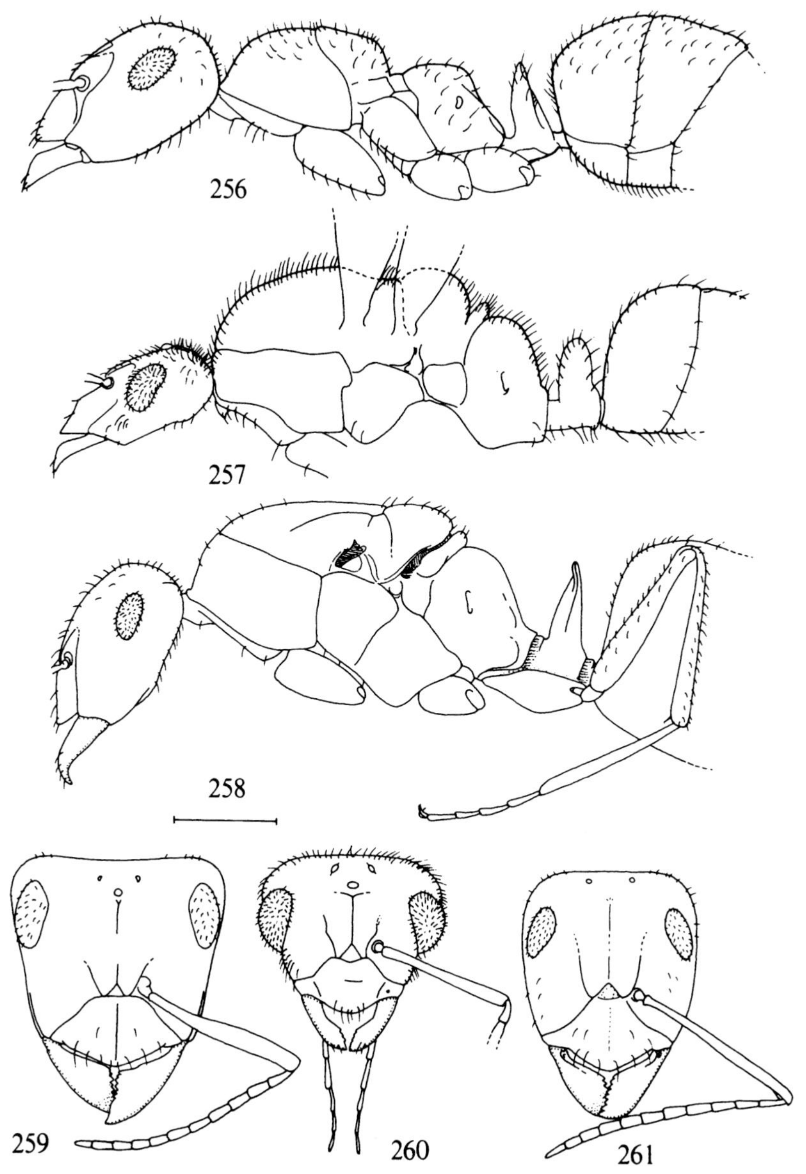

Figs. 256–261. *Formica pratensis* Retzius. – 256: worker in profile; 257: male in profile; 258: queen in profile; 259: head of queen in dorsal view; 260: head of male in dorsal view; 261: head of worker in dorsal view. Scale: 1 mm.

demarcate the extreme forms as a separate species but samples mainly from coastal areas and offshore islands in Nylandia include some extremely hairy specimens with queens consistently having wide spaced micropunctures which are well outside the range of *F. lugubris* as described by Yarrow (1955) and Betrem (1960). Bondroit (1917) briefly described a form, *F. rufa* var. *nylanderi*, as having long outstanding body and antennal hairs and *F. nylanderi* could be a suitable name for this form, if distinguished as a species.

F. lugubris spreads by colony fission but also by the adoption of fertile queens by *Formica lemani*. Such mixed incipient nests often under stones have frequently been seen in Norway and North Sweden (Collingwood, 1959).

62. ***Formica pratensis*** Retzius, 1783
Figs. 170, 175, 256–262.

Formica pratensis Retzius, 1783: 75; Betrem, 1965.
Formica nigricans Emery; Yarrow, 1955; Betrem, 1960.

Worker. Bicoloured with gaster, occiput and frons matt black, not shining; gaster more or less thickly pubescent. Black patch on promesonotum variable but in typical specimens clearly demarcated. Eyes thickly haired; occiput with short to medium length fringing hairs, sometimes reduced to very few. Antennal scapes without protruding hairs. Femora and tibiae fringed with hairs on extensor surfaces. Length: 4.5–9.5 mm.

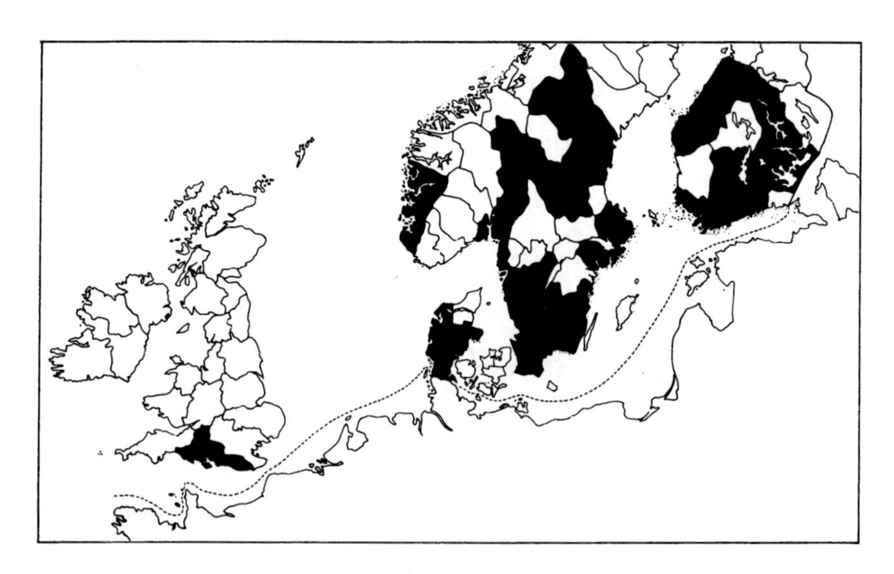

Fig. 262. Distribution of *Formica pratensis* Retzius, a southern boreal species.

Queen. As worker with all dark areas pubescent, closely sculptured and matt. Basal face of gaster and occiput with short hairs, sometimes difficult to discern or absent. Length: 9.5–11.3 mm.

Male. Matt black with pubescent gaster. Suberect hairs present on dorsum of all gaster tergites and extensor surfaces of femora and tibiae. Eye and outstanding genal hairs below eyes always plentiful and prominent. Length: 9.5–11.5 mm.

Distribution: Locally common in Denmark (only recorded from Jutland) and Southern Fennoscandia to latitude 63°. – Rare in South England. – Range: Portugal to Siberia, North Italy to Central Sweden.

Biology. This is the black backed meadow ant characteristic of rough alpine pastures but also common on woodland borders and scrubby heathland throughout lowland Europe and South Fennoscandia. Colonies are isolated single nests with one or very few queens. Jensen (1977) gives population estimates for this species in Denmark of up to 60,000 workers. Nests are smaller than with *F. rufa* and other species of this group and nest materials are coarser. A morphologically indistinguishable form '*pratensoides*' Gösswald (1951), which is polygynous with many grouped nests, occurs locally in Germany and the Netherlands, often in shaded woodland, but has not been recorded from Denmark or Fennoscandia. Brood development begins later in the spring with sexuals normally appearing in July.

63. ***Formica nigricans*** Emery, 1909
Figs. 263–266.

Formica rufa pratensis var. *nigricans* Emery, 1909: 187.
Formica cordieri Bondroit, 1918: 126.
Formica nigricans Emery; Betrem, 1965.

Worker. Similar to *F. pratensis* in all respects except that body and appendage hairs tend to be more abundant and longer and most samples have 2 or 3 subdecumbent hairs protruding from the upper surface of the scape. Length: 4.5–9.5 mm.

Queen. Similar to *F. pratensis* in colour but with long bent hairs arising from propodeum scale and basal face of gaster in addition to other parts of body which is often extremely hairy. Scapes often, and tibiae normally, with long suberect hairs. Length: 10.0–11.0 mm.

Male. Pubescence and appearance as *F. pratensis* but appendage hairs thicker and longer, with occasional erect hairs on hind tibiae longer than half width of tibiae. Pubescence on scape merging into subdecumbent protruding short hairs. Length: 9.5–10.5 mm.

Distribution. Very local. Denmark: SJ, EJ, WJ. – Sweden: from south to Vrm. – Norway: HO. – Range: Central Italy to Central Sweden, Portugal to Central Asia.

Biology. This species occurs in isolated nests and sometimes in a loose group of nests. In appearance and behaviour it is similar to *F. pratensis* but has been found nesting in dry sheltered banks, open lowland woodland and among scrub in partial shade in Mediterranean areas.

Note. Dlussky (1967) doubted whether *F. nigricans* could be specifically distinct and Paraschivescu (1972) gave evidence to suggest that the two forms intergraded in pilosity characters. The strongest argument for their separate identity lies in their geographical range. *F. nigricans* occurs much further to the south in Italy than *F. pratensis*, is characteristic of the Mediterranean area where *F. pratensis* has not been recorded. Conversely all samples from England, Finland, Channel Islands and most of the Netherlands are *F. pratensis* with no overlap in morphology. One aberrant polygynous polycalic colony is known from the southwest Netherlands with queens of mainly *F. nigricans* type but with variable pilosity.

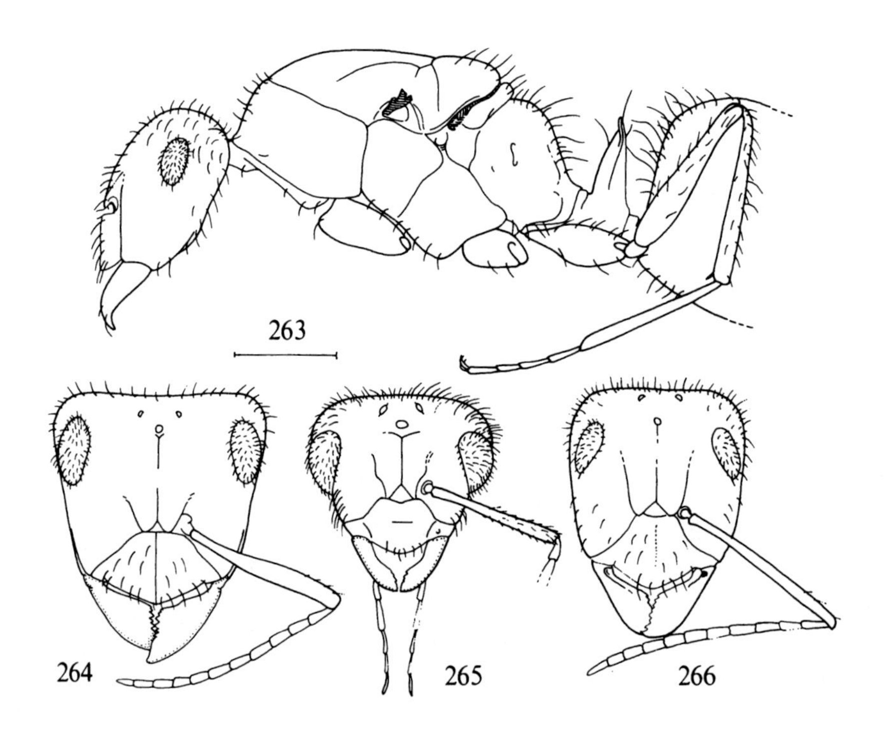

Figs. 263–266. *Formica nigricans* Emery. – 263: queen in profile; 264: head of queen in dorsal view; 265: head of male in dorsal view; 266: head of worker in dorsal view. Scale: 1 mm.

Genus *Polyergus* Latreille, 1805

Polyergus Latreille, 1805: 256.
Type-species: *Formica rufescens* Latreille, 1798.

This genus includes three palaearctic and several nearctic species immediately recognisable from other formicine ants by the long sickle shaped toothless mandibles and much reduced thin palps, formula 4: 2. Members of this genus raid nests of ants of the *Formica fusca* group to carry away pupae to be reared as auxiliaries in the home nest. *Polyergus* workers are unable to feed themselves and are entirely dependent on captive *Formica* for their survival.

64. ***Polyergus rufescens*** (Latreille, 1798)
Figs. 267, 268.

Formica rufescens Latreille, 1798: 44.

Worker. Reddish brown to brownish; head elongate; clypeus foreshortened with flat anterior border, armed with long edentate sickle shaped mandibles. Maxillary palps short and very slender. Scale nodal; propodeum sharply raised in profile. Gula, clypeus, occiput, dorsum of alitrunk with erect hairs, more profuse on gaster. Length: 5.0–7.0 mm.

Ergatoid queen. As worker but with more massive alitrunk and hairs more sparse on alitrunk and gaster. Length: 7.0–9.0 mm.

Normal queen. Winged or with normal sclerites and normal queen shaped broad alitrunk. Colour as worker but appearance more shining with body hairs more sparse. Length: 8.0–9.5 mm.

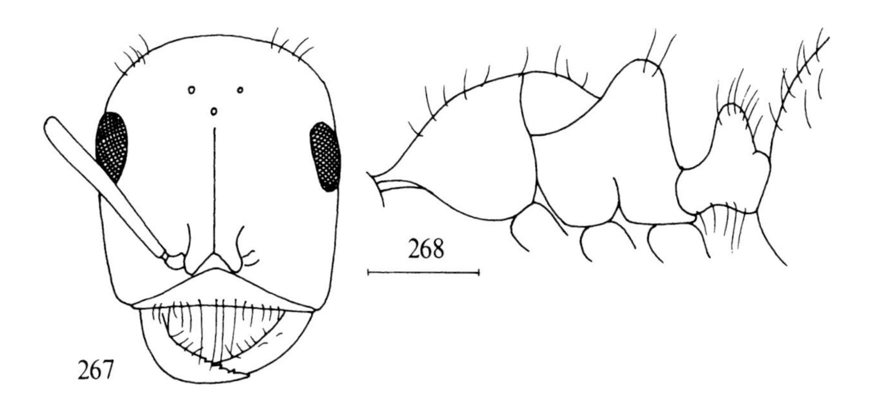

Figs. 267, 268. *Polyergus rufescens* (Latr.). – 267: head of worker in dorsal view; 268: alitrunk of worker in profile. Scale: 1 mm.

Male. Black with eyes, legs and funiculi pale. Mandibles edentate, very thin and reduced. Palps very reduced. Head short and broad relative to alitrunk; eyes very large, hairless. Scale nodal but thinner in profile than in female castes, emarginate in front view. Scale and gaster with abundant long hairs. Length: 6.0–7.0 mm.

Distribution. South and Central Sweden, very local, recorded from Sk., Bl., Hall., Sm., Öl., Upl. – Not found in British Isles. – Range: Spain to South Russia, Central Italy to Stockholm; rarein Belgium and Netherlands.

Biology. This is the famous Amazon ant. Raiding forays are carried out by small troops of workers on hot days after mid summer and are invariably successful. The adversary ants are decapitated or have their heads pierced by the *Polyergus* with their pincerlike jaws. The worker pupae of members of the *Formica fusca* group, usually *F. fusca* or *F. rufibarbis*, are carried back to the parent nest for rearing as auxiliary workers by ants of the same or similar species introduced by earlier raids. Single queens whether ergatoid or normal secure adoption in nests of the auxiliary species by destroying and replacing the host queen. A mature colony may consist of a few dozen to a few hundred *Polyergus* workers and many more up to a few thousand of the auxiliary species. Queens and workers are unable to feed themselves or to rear brood. Nests are under flat stones in warm sheltered places.

Catalogue

				DENMARK												
		Germany	G. Britain	SJ	EJ	WJ	NWJ	NEJ	F	LFM	SZ	NWZ	NEZ	B	Sk.	Bl.
Hypoponera punctatissima (Roger)	1	●	●		●				●	●			●		●	
Ponera coarctata (Latr.)			●													
Tapinoma erraticum (Latr.)	2	●	●													
Myrmica gallieni Bondr.	3	●								●						
M. lobicornis Nyl.	4	●	●	●	●	●	●	●					●		●	●
M. rubra (L.)	5	●	●	●	●	●	●	●	●	●	●				●	●
M. ruginodis Nyl.	6	●	●	●	●	●	●	●	●	●	●		●	●	●	●
M. rugulosa Nyl.	7	●		●	●	●	●	●	●			●	●		●	
M. sabuleti Mein.	8	●	●	●	●	●	●	●		●					●	●
M. scabrinodis Nyl.	9	●	●	●	●	●	●	●	●	●	●		●		●	●
M. schencki Emery	10	●	●	●	●	●	●	●					●	●	●	●
M. specioides Bondr.	11	●	●							●						
M. sulcinodis Nyl.	12	●	●			●	●	●							●	
M. hirsuta Elmes			●													
Sifolinia karavajevi (Arnoldi)	13		●												●	
Stenamma westwoodii Westwood	14	●	●								●				●	●
Diplorhoptrum fugax (Latr.)	15	●	●												●	
Myrmecina graminicola (Latr.)	16	●	●		●										●	
Leptothorax acervorum (Fabr.)	17	●	●		●	●	●	●	●	●		●	●	●	●	●
L. muscorum (Nyl.)	18	●			●	●									●	●
L. nylanderi (Förster)	19	●	●												●	●
L. corticalis (Schenck)	20	●														
L. interruptus (Schenck)	21	●	●													
L. tuberum (Fabr.)	22	●	●					●						●	●	●
L. unifasciatus (Latr.)	23		●													
Formicoxenus nitidulus (Nyl.)	24	●	●		●				●				●	●	●	
Harpagoxenus sublaevis (Nyl.)	25	●			●	●	●								●	
Anergates atratulus (Schenck)	26	●	●										●	●		
Strongylognathus testaceus (Schenck)	27	●	●													
Tetramorium caespitum (L.)	28	●	●	●	●	●	●	●	●	●	●	●	●	●	●	●
Camponotus fallax (Nyl.)	29	●														
C. vagus (Scop.)	30															
C. herculeanus (L.)	31	●		●	●		●	●					●		●	●
C. ligniperda (Latr.)	32	●			●					●				●	●	●
Lasius flavus (Fabr.)	33	●	●	●	●	●	●	●	●	●	●	●	●	●	●	●
L. alienus (Förster)	34	●	●	●	●	●		●	●	●		●	●		●	●
L. brunneus (Latr.)	35	●	●							●						●
L. niger (L.)	36	●	●	●	●	●	●	●	●	●	●	●	●	●	●	●

SWEDEN

	Hall.	Sm.	Öl.	Gtl.	G. Sand.	Ög.	Vg.	Boh.	Dlsl.	Nrk.	Sdm.	Upl.	Vstm.	Vrm.	Dlr.	Gstr.	Hls.	Med.	Hrj.	Jmt.	Ång.	Vb.	Nb.	Ås. Lpm.	Ly. Lpm.	P. Lpm.	Lu. Lpm.	T. Lpm.
1												●			●													
2			●	●																								
3				●																								
4	●	●	●	●		●	●	●	●	●	●	●	●	●	●		●	●	●	●	●	●	●	●	●		●	●
5	●	●	●	●	●	●	●	●	●	●	●	●	●	●	●	●	●	●	●	●	●	●	●					
6	●	●	●	●	●	●	●	●	●	●	●	●	●	●	●	●	●	●	●	●	●	●	●	●	●	●	●	●
7	●	●	●	●		●	●		●	●																		
8	●	●	●	●		●	●	●	●	●	●	●	●	●														
9	●	●	●	●		●	●	●	●	●	●	●	●	●	●	●	●	●	●	●	●	●	●	●	●		●	
10	●	●	●	●	●	●	●	●		●	●	●	●		●													
11																												
12	●	●	●			●	●	●	●	●	●	●	●	●	●		●	●	●	●	●	●	●	●	●	●	●	●
13																												
14		●		●			●																					
15			●	●																								
16		●	●	●			●																					
17	●	●	●	●	●	●	●	●	●	●	●	●	●	●	●	●	●	●	●	●	●	●	●	●	●	●	●	●
18		●	●		●	●	●	●	●	●	●	●	●	●	●		●	●										
19		●		●	●	●	●		●																			
20			●									●	●															
21				●	●																							
22	●	●	●	●	●	●	●	●			●	●				●		●										
23			●																									
24	●	●	●	●		●	●		●	●		●			●		●	●		●			●					
25		●	●			●			●	●	●	●	●	●	●			●		●			●	●	●	●		●
26	●		●	●		●																						
27		●	●																									
28	●	●	●	●	●	●	●	●	●	●	●	●						●										
29													●															
30			●	●																								
31	●	●	●	●	●	●	●	●	●	●	●	●	●	●	●	●	●	●	●	●	●	●	●	●	●	●	●	●
32	●	●	●	●	●	●	●	●	●		●	●	●			●		●										
33	●	●	●	●		●	●	●	●	●	●	●	●	●	●	●	●	●		●	●	●	●		●			
34	●	●	●	●	●	●	●	●		●	●	●			●			●										
35		●	●	●		●					●	●	●		●													
36	●	●	●	●	●	●	●	●	●	●	●	●	●	●	●	●	●	●		●	●	●	●					

		NORWAY														
		Ø + AK	HE (s + n)	O (s + n)	B (ø + v)	VE	TE (y + i)	AA (y + i)	VA (y + i)	R (y + i)	HO (y + i)	SF (y + i)	MR (y + i)	ST (y + i)	NT (y + i)	Ns (y + i)
Hypoponera punctatissima (Roger)	1	●														
Ponera coarctata (Latr.)																
Tapinoma erraticum (Latr.)	2															
Myrmica gallieni Bondr.	3															
M. lobicornis Nyl.	4	●	●	●	●	●		●	●	●	●			●	●	●
M. rubra (L.)	5	●	●	●	●	●	●		●	●	●			●	●	●
M. ruginodis Nyl.	6	●	●	●	●	●	●	●	●	●	●	●	●	●	●	●
M. rugulosa Nyl.	7															
M. sabuleti Mein.	8	●	●			●				●	●	●				
M. scabrinodis Nyl.	9	●	●	●	●	●		●	●	●	●			●		●
M. schencki Emery	10	●				●				●						
M. specioides Bondr.	11															
M. sulcinodis Nyl.	12	●	●	●	●	●		●	●	●	●		●	●		●
M. hirsuta Elmes																
Sifolinia karavajevi (Arnoldi)	13		●													
Stenamma westwoodii Westwood	14								●		●					
Diplorhoptrum fugax (Latr.)	15															
Myrmecina graminicola (Latr.)	16															
Leptothorax acervorum (Fabr.)	17	●	●	●	●	●	●	●	●	●	●	●	●	●	●	●
L. muscorum (Nyl.)	18	●	●	●	●	●				●	●					
L. nylanderi (Förster)	19															
L. corticalis (Schenck)	20															
L. interruptus (Schenck)	21															
L. tuberum (Fabr.)	22	●	●	●	●	●	●	●								
L. unifasciatus (Latr.)	23															
Formicoxenus nitidulus (Nyl.)	24	●		●	●	●		●	●	●						●
Harpagoxenus sublaevis (Nyl.)	25	●			●	●			●	●	●					
Anergates atratulus (Schenck)	26															
Strongylognathus testaceus (Schenck)	27															
Tetramorium caespitum (L.)	28	●	●		●	●	●	●	●	●	●		●			
Camponotus fallax (Nyl.)	29															
C. vagus (Scop.)	30															
C. herculeanus (L.)	31	●	●	●	●	●		●	●	●	●	●	●	●	●	●
C. ligniperda (Latr.)	32	●	●		●	●	●	●	●	●		●	●			
Lasius flavus (Fabr.)	33	●	●	●	●	●	●	●	●	●	●	●	●	●		
L. alienus (Förster)	34	●	●			●										
L. brunneus (Latr.)	35	●			●	●		●								
L. niger (L.)	36	●	●	●	●	●	●	●	●	●	●	●				

	Nn (ø + v)	TR (y + i)	F (v + i)	F (n + ø)	Al	Ab	N	Ka	St	Ta	Sa	Oa	Tb	Sb	Kb	Om	Ok	Ob S	Ob N	Ks	LkW	LkE	Le	Li	Vib	Kr	Lr
					FINLAND																				USSR		
1						●	●			●			●														
2																											
3							●																				
4	●	●	●	●	●	●	●	●	●	●	●	●	●	●	●	●	●	●	●	●	●	●	●	●	●	●	
5	●			●	●	●	●	●	●	●	●	●	●	●	●	●	●	●	●	●					●	●	
6	●	●	●	●	●	●	●	●	●	●	●	●	●	●	●	●	●	●	●	●	●	●	●	●	●	●	●
7						●	●	●		●				●													
8						●	●			●	●															●	
9	●				●	●	●	●	●	●	●	●	●	●	●	●	●	●	●	●	●				●	●	
10					●	●	●			●	●															●	
11						●	●																				
12	●			●	●	●	●	●		●		●	●	●	●	●	●	●	●	●	●	●		●		●	●
13											●																
14																											
15																											
16																											
17	●	●	●	●	●	●	●	●	●	●	●	●	●	●	●	●	●	●	●	●	●	●	●	●	●	●	●
18					●	●	●	●	●	●	●			●		●											
19																											
20																											
21																											
22					●	●	●	●			●															●	
23																											
24	●	●		●	●	●	●	●	●	●	●	●	●	●			●	●	●	●	●	●			●	●	
25			●			●	●			●		●	●	●			●	●	●	●	●	●				●	●
26																											
27																											
28					●	●	●	●	●	●	●	●	●	●													
29																											
30						●		●																		●	
31	●	●	●	●	●	●	●	●	●	●	●	●	●	●	●	●	●	●	●	●	●	●	●	●	●	●	●
32					●	●	●	●		●	●														●	●	
33				●	●	●	●	●	●	●	●	●	●	●	●	●	●	●	●	●					●	●	
34						●	●	●	●	●					●			●	●								
35																											
36					●	●	●	●	●	●	●	●	●	●	●	●	●	●	●	●					●	●	

				DENMARK												
		Germany	G. Britain	SJ	EJ	WJ	NWJ	NEJ	F	LFM	SZ	NWZ	NEZ	B	Sk.	Bl.
L. emarginatus (Oliv.)		●	●													
L. fuliginosus (Latr.)	37	●	●	●	●	●	●	●	●	●	●	●	●	●	●	●
L. umbratus (Nyl.)	38	●	●	●	●	●		●	●	●			●		●	●
L. meridionalis (Bondr.)	39	●	●		●		●	●		●			●	●	●	
L. bicornis (Förster)	40															
L. mixtus (Nyl.)	41	●	●		●				●				●		●	●
L. carniolicus (Mayr)	42															
Plagiolepis vindobonensis Lomnicki			●													
Formica fusca L.	43	●	●	●	●	●	●	●	●	●			●		●	●
F. gagatoides Ruzsky	44															
F. lemani Bondr.	45	●	●													
F. transkaucasica Nasonov	46	●	●	●	●	●		●				●	●		●	
F. cinerea Mayr	47	●		●	●	●	●	●							●	●
F. cunicularia Latr.	48	●	●		●										●	●
F. rufibarbis Fabr.	49	●	●		●	●	●	●	●	●		●	●	●	●	●
F. exsecta Nyl.	50	●	●	●	●	●	●	●	●				●		●	●
F. foreli Emery	51												●			
F. forsslundi Lohmander	52				●	●									●	
F. pressilabris Nyl.	53	●		●	●	●	●	●		●		●	●		●	●
F. suecica Adlerz	54															
F. uralensis Ruzsky	55	●		●	●	●		●							●	
F. sanguinea Latr.	56	●	●		●	●	●	●	●			●	●		●	●
F. truncorum Fabr.	57	●			●			●		●	●	●	●	●	●	●
F. rufa L.	58	●	●	●	●	●	●	●	●			●	●		●	●
F. polyctena Förster	59	●			●			●					●		●	●
F. aquilonia Yarrow	60	●	●		●											
F. lugubris Zett.	61	●	●													
F. pratensis Retzius	62	●	●	●	●	●	●								●	●
F. nigricans Emery	63	●		●	●	●									●	●
Polyergus rufescens (Latr.)	64	●													●	●

	Hall.	Sm.	Öl.	Gtl.	G. Sand.	Ög.	Vg.	Boh.	Dlsl.	Nrk.	Sdm.	Upl.	Vstm.	Vrm.	Dlr.	Gstr.	Hls.	Med.	Hrj.	Jmt.	Ång.	Vb.	Nb.	Ås. Lpm.	Ly. Lpm.	P. Lpm.	Lu. Lpm.	T. Lpm.
37	●	●	●	●		●	●	●	●	●	●	●	●	●	●			●										
38	●	●	●	●		●	●		●	●	●	●	●	●	●			●										
39	●		●												●													
40		●																										
41	●	●	●	●	●	●	●	●	●	●	●	●	●	●	●			●										
42		●	●	●																								
43	●	●	●	●	●	●	●	●	●	●	●	●	●	●	●	●	●	●	●	●	●	●			●			
44															●			●	●	●		●	●		●	●	●	●
45		●				●	●			●			●	●	●	●	●		●	●	●	●	●	●	●	●	●	●
46	●	●				●	●			●	●	●	●	●	●	●		●	●	●		●	●	●	●			●
47	●		●										●	●	●													
48	●		●	●	●																							
49	●	●	●	●	●	●	●	●		●	●	●	●		●	●												
50	●	●	●	●		●	●	●	●	●	●	●	●	●	●	●	●	●	●	●	●	●	●	●	●	●	●	●
51																												
52	●	●				●	●		●	●		●		●	●				●									
53	●	●	●			●	●	●	●	●		●		●	●	●		●										
54		●					●			●			●	●	●		●	●	●	●	●	●	●	●	●		●	
55	●	●				●	●			●	●	●		●	●					●	●	●	●	●	●			●
56	●	●	●	●	●	●	●	●	●	●	●	●	●	●	●	●	●	●	●	●	●	●	●	●	●	●	●	
57	●	●	●	●	●	●	●	●	●	●	●	●	●	●	●	●	●	●	●	●	●	●	●	●	●	●	●	●
58	●	●	●	●	●	●	●	●	●	●	●	●	●	●	●	●	●	●		●								
59	●	●	●	●		●	●			●	●	●		●														
60		●				●			●				●		●		●	●	●	●	●	●	●	●	●	●	●	●
61		●				●							●	●	●		●	●	●		●	●	●	●		●	●	●
62	●	●				●	●	●			●	●			●		●	●		●								
63	●	●	●			●		●		●	●	●		●														
64	●	●	●									●																

		NORWAY														
		Ø + AK	HE (s + n)	O (s + n)	B (ø + v)	VE	TE (y + i)	AA (y + i)	VA (y + i)	R (y + i)	HO (y + i)	SF (y + i)	MR (y + i)	ST (y + i)	NT (y + i)	Ns (y + i)
L. emarginatus (Oliv.)																
L. fuliginosus (Latr.)	37	●	●	●	●	●	●	●								
L. umbratus (Nyl.)	38	●	●	●	●	●										
L. meridionalis (Bondr.)	39					●										
L. bicornis (Förster)	40															
L. mixtus (Nyl.)	41	●	●		●		●			●	●					
L. carniolicus (Mayr)	42															
Plagiolepis vindobonensis Lomnicki																
Formica fusca L.	43	●	●	●		●	●			●	●	●				
F. gagatoides Ruzsky	44			●	●					●	●	●	●	●		●
F. lemani Bondr.	45	●	●	●	●	●	●	●	●	●	●	●	●	●	●	●
F. transkaucasica Nasonov	46		●			●										
F. cinerea Mayr	47		●													
F. cunicularia Latr.	48															
F. rufibarbis Fabr.	49	●			●	●		●	●							
F. exsecta Nyl.	50	●	●	●	●	●		●	●	●	●		●	●	●	●
F. foreli Emery	51															
F. forsslundi Lohmander	52		●													
F. pressilabris Nyl.	53	●	●			●										
F. suecica Adlerz	54		●			●				●		●				
F. uralensis Ruzsky	55		●													
F. sanguinea Latr.	56	●	●	●	●	●			●	●	●					
F. truncorum Fabr.	57	●		●	●	●				●	●	●	●		●	
F. rufa L.	58	●	●	●	●	●		●	●		●	●				
F. polyctena Förster	59	●				●										
F. aquilonia Yarrow	60	●	●	●	●	●	●	●	●	●	●	●	●	●	●	●
F. lugubris Zett.	61	●	●	●	●	●		●	●	●	●	●	●	●	●	●
F. pratensis Retzius	62	●	●			●				●	●					
F. nigricans Emery	63										●					
Polyergus rufescens (Latr.)	64															

					FINLAND																				USSR		
	Nn (ø + v)	TR (y + i)	F (v + i)	F (n + ø)	Al	Ab	N	Ka	St	Ta	Sa	Öa	Tb	Sb	Kb	Om	Ok	Ob S	Ob N	Ks	LkW	LkE	Le	Li	Vib	Kr	Lr
37					●	●	●	●	●	●	●	●	●	●	●	●	●										
38					●	●	●	●		●																●	
39							●																				
40																											
41						●	●	●		●	●															●	
42																											
43						●	●	●	●	●	●	●	●	●	●	●		●	●	●					●	●	
44		●	●	●										●					●	●	●	●	●	●		●	●
45	●	●	●	●	●	●	●	●	●	●	●	●	●	●	●	●	●	●	●	●	●	●	●	●		●	●
46						●	●	●	●	●	●	●	●	●	●	●	●	●	●	●					●	●	
47						●	●				●			●	●	●	●	●							●		
48																											
49					●	●	●	●		●	●														●	●	
50	●		●	●	●	●	●	●	●	●	●	●	●	●	●	●	●	●	●	●	●	●	●	●	●	●	●
51																											
52						●			●				●						●						●	●	
53					●		●			●	●			●			●		●	●							
54						●	●									●		●	●								
55				●	●	●	●	●	●	●	●	●	●	●	●	●	●	●	●							●	●
56				●	●	●	●	●	●	●	●	●	●	●	●	●	●	●	●	●	●	●	●	●	●	●	●
57		●	●	●	●	●	●	●	●	●	●		●	●	●	●		●	●	●	●	●		●	●	●	●
58					●	●	●	●	●	●	●	●	●	●		●									●		
59							●	●																	●		
60	●	●	●	●	●	●	●	●	●	●	●	●	●	●	●	●	●	●	●	●	●	●	●	●		●	
61	●	●	●		●	●	●	●		●		●	●	●		●		●	●	●	●	●	●	●		●	
62						●	●			●	●	●		●		●											
63																											
64																											

Literature

Adlerz, C., 1902: Myrmecologiska Studier. IV. *Formica suecica* n. sp. Eine neue schwedische Ameise. – Öfvers.K.VetenskAkad. Förh. 59 (8): 263–265.

Adlung, K.G., 1966: A critical evaluation of the European research on use of Red Wood Ants (*Formica rufa* Group) for the protection of forest against harmful insects. – Z.angew.Ent. 57: 167–189.

Arnoldi, K. V., 1930: Studien über die Systematik der Ameisen. VI. – Zool.Anz.91: 267–283.

– 1934: Vorläufige Ergebnisse einer biometrischen Untersuchen einiger *Myrmica*-Arten aus dem Europäischen Teile der USSR. – Folia zool.hydrobiol. 6 (2): 151–174.

Baroni Urbani, C., 1968: Über die eigenartige Morphologie der männlichen Genitalien des Genus *Diplorhoptrum* Mayr und die taxonomischen Schlussfolgerungen. – Z.Morph. Ökol. Tiere 63: 63–74.

– 1971: Catalogo delle Specie di Formicidae d'Italia. – Memorie Soc.ent.ital. 50: 5–287.

Baroni Urbani, C. & Collingwood, C.A., 1977: The Zoogeography of ants (Hymenoptera, Formicidae) in Northern Europe. – Acta zool.fenn. 152: 1–34.

Bernard, F., 1968: Les Fourmis d'Europe occidentale et septentrionale. – Faune de l'Europe et du Bassin Mediterranéen 3. Paris.

Betrem, J. G., 1960: Ueber die Systematik der *Formica rufa*-Gruppe. – Tijdschr.Ent. 103: 51–81.

– 1964: Einiger Bemerkungen über *Formica* Material aus Nordost Karelien. – Waldhygiene 5: 215–216.

– 1965: Über die Systematik einiger Arten der Gruppe *Formica rufa*. – Collana Verde 16: 80–85.

Bisgaard, Ch., 1944: Meddelelse om nogle nye Myrer for Danmarks Fauna. – Ent.Meddr 24: 117–126.

Blum, M.S., 1969: Alarm Pheromones. – A.Rev.Ent. 14: 57–80.

– 1970: The chemical basis of insect sociality. *In:* Chemicals controlling insect behaviour, pp. 61–94. Acad.Press, New York.

Bolton, B., 1976: The ant tribe Tetramoriini (Hymenoptera: Formicidae) – Bull.Br.Mus. nat.Hist. (Ent.) 34 (5): 281–379.

– 1977: The ant tribe Tetramoriini (Hymenoptera: Formicidae). The genus *Tetramorium* Mayr in the Oriental and Indo-Australian regions and in Australia. – Bull.Br.Mus.nat. Hist. (Ent.) 36 (2): 67–151.

Bolton, B. & Collingwood, C.A., 1975: Hymenoptera, Formicidae – Handbk Ident.Br. Insects VI (3c): 1–34.

Bondroit, J., 1917: Diagnoses de trois nouveaux *Formica* d'Europe. – Bull.Soc.ent.Fr., 1917/18: 186–188.

– 1918: Les fourmis de France et de Belgique. – Annls Soc.ent.Fr. 87: 1–174.

– 1919: Notes diverses sur les fourmis d'Europe. – Annls Soc.ent.Belg. 59: 143–158.

– 1920: Supplément aux fourmis de France et de Belgique. – Annls Soc.Ent.Fr. 88: 299–305.

Bourne, R.A., 1973: A taxanomic study of the ant genus *Lasius* Fabricius in the British Isles (Hymenoptera, Formicidae). – J.Ent. (B) 42: 17–27.

Boven, J.K.A. van, 1959: Mieren (Formicidae). - Wet.Meded.K.ned.natuurh.Veren. No. 30 (VI): 1-52.
- 1977: De Mierenfauna van België (Hymenoptera: Formicidae). - Acta zool.path.Antv. 67: 1-191.
Brian, M.V. & Brian, A.D., 1949: Observations on the taxonomy of the ants *Myrmica rubra* L. and *M.laevinodis* Nylander (Hymenoptera: Formicidae). - Trans.R.ent.Soc.Lond. 100: 393-409.
- 1955: On the forms *macrogyna* and *microgyna* of the ant *Myrmica rubra* L. - Evolution 9: 280-290.
Brian, M.V., 1965: Social Insect Populations. - Academic Press, London, New York. 135 pp.
- 1977: Ants. - Collins: New Naturalist, London. 223 pp.
Buschinger, A., 1966: *Leptothorax (Mychothorax) muscorum* (Nylander) and *Leptothorax (M.) gredleri* Mayr zwei gute Arten. - Insectes soc. 13: 165-172.
Collingwood, C.A. 1958: The ants of the genus *Myrmica* in Britain. - Proc.R.ent.Soc.Lond. (A) 33: 65-75.
- 1959: Scandinavian Ants. - Entomologist's Rec.J.Var. 71: 77-81.
- 1961: Ants in Finland. - Ibid. 73: 190-195.
- 1963a: Three species of ants new to Norway. - Ibid. 75: 225-228.
- 1963b: The *Lasius (Chthonolasius) umbratus* (Hym., Formicidae) complex in North Europe. - Entomologist 96: 145-158.
- 1971: A synopsis of the Formicidae of North Europe. - Entomologist 104: 150-176.
- 1974: A revised list of Norwegian ants (Hymenoptera, Formicidae). - Norsk ent. Tidsskr. 21: 31-35.
- 1976: Mire invertebrate fauna at Eidskog, Norway. III. Formicidae (Hymenoptera Aculeata). - Norw.J.Ent. 23: 185-187.
Curtis, J., 1829: British Entomology 6, London.
Czechowski, W., 1975: Bionomics of *Formica (Coptoformica) pressilabris* (Hymenoptera, Formicidae). - Annls zool. 33: 103-125.
Donisthorpe, H., 1927: Guests of British Ants, their habits and life histories. - Routledge, London. 244 pp.
Douwes, P., 1976a: Interessante fynd av myror (Hym., Formicidae). - Entomologen 5: 25.
- 1976b: Sveriges myror - illustrerade bestämningstabeller över arbetarna. - Ibid.: 37-54.
- 1977: *Sifolinia karavajevi*, en för Sverige ny myra (Hym., Formicidae), - Ent.Tidsskr. 98: 147-148.
Dlussky, G. M., 1967: Ants of the genus *Formica*. - Izdatelstvo "Nauka", Moscow, 237 pp. (In Russian).
Eichler, W., 1976: Distribution, spread and hygienic importance of the pharaoh's ants in Europe. - Proc. II Internat. Symposium, IUSSI, Warsaw: 19-20.
Elmes, G.W., 1973a: Miniature queens of the ant *Myrmica rubra* L. (Hymenoptera, Formicidae). - Entomologist 106: 133-136.
- 1973b: Observations on the density of queens in natural colonies of *Myrmica rubra* L. (Hymenoptera: Formicidae). - J.Anim. Ecol. 42: 761-771.
- 1978: A morphometric comparison of three closely related species of *Myrmica* (Formicidae), including a new species from England. - Syst.Ent. 3: 131-145.
Emery, C., 1895: Beiträge zur Kenntnis der nordamerikanischen Ameisenfauna. - Zool.Jb. (Syst.) 8: 257-360.
- 1907: Intorno all'origine delle formiche dulotiche, parassite e mirmecophile. - Rc.Sess. Acad.Sci.Ist.Bologna 1908-09: 36-51.

- 1909: Beiträge zur Monographie der Formiciden des palaearktischen Faunengebietes. VII. *Formica*. - Dt.ent.Z. 1909: 179–204.
- 1916: Formiche d'Italia nouve o critiche. - Rc.Sess.Accad.Sci. Ist.Bologna 1915–16: 53–66.
Ettershank, G., 1966: A generic revision of the world Myrmicinae related to *Solenopsis* and *Pheidologeton*. - Aust.J.Zool. 14: 73–171.
Faber, W., 1967: Beiträge zur Kenntnis sozialparasitischer Ameisen. 1. *Lasius (Austrolasius)* n.sg. *reginae* n.sp. eine neue temporär sozialparasitische Erdameise aus Österreich. - Pflanzenschutz-berichte 36: 73–108.
Fabricius, J.C., 1775: Systema entomologiae. 832 pp. Flensburgi et Lipsiae.
- 1781: Species insectorum. Hamburgi et Kilonii.
- 1793: Entomologia systematica emendata et aucta. Vol. 2, 519 pp. Hafniae.
- 1804: Systema piezatorum. 440 pp. Brunsvigae.
Finzi, B., 1926: La forma europea del genere *Myrmica* Latr. - Boll.Soc.adriat.Sci.nat. 29: 71–119.
Fjellberg, A: 1975: Occurrence of *Formica uralensis* Ruzsky (Hymenoptera, Formicidae) in Pasvik, North Norway. - Norw.J.Ent. 22: 83.
Forel, A., 1874: Les fourmis de la Suisse. - Neue Denkschr.allg.schweiz.Ges.ges.Naturw., 26. Zürich.
- 1893: Sur la classification de la famille des Formicides, avec remarques synonymiques. - Annls Soc.ent.Belg. 37: 161–167.
- 1915: Fauna insectorum Helvetiae - Hymenoptera, Formicidae. Die Ameisen der Schweiz. - Mitt.schweiz.ent.Ges., Beilage zu Heft 7/8, vol. 12, 1–77.
Forsslund, K.-H., 1947: Svenska myror, 1–10. - Ent.Tidskr. 68: 67–80.
- 1949: Svenska myror, 11–14. - Ibid. 70: 19–32.
- 1957a: Svenska myror, 15–19. - Ibid. 78: 33–40.
- 1957b: Catalogus Insectorum Sueciae. Hymenoptera: Fam. Formicidae. - Opusc.ent. 22: 70–78.
Förster, A., 1850: Hymenopterologische Studien - Formicariae. 74 pp. Aachen.
Gallé, L., 1973: Thermoregulation in the nest of *Formica pratensis* Retz. (Hymenoptera: Formicidae). - Acta biol. Szeged 19: 139–143.
Gösswald, K., 1941: Rassenstudien an der Roten Waldameise, *Formica rufa* L., auf systematischer, physiologischer und biologischer Grundlage. - Z. angew. Ent. 28: 62–124.
- 1951: Zur Biologie, Ökologie und Morphologie einer neuen Varietät der kleinen roten Waldameise, *Formica minor pratensoides*. - Ibid. 32: 433–457.
Holgersen, H., 1942: Ants of Northern Norway. - Tromsø Mus. Årsh.Naturhistorisk Avd. no. 24, = vol. 63 (2): 1–33.
- 1943: *Formica gagatoides* Ruzsky in Norway. - Ibid. no. 26, = vol. 64. (1): 1–17.
- 1944: The Ants of Norway. - Nyt Mag.Naturv. 84: 163–202.
Hölldobler, B., 1960: Über die Ameisenfauna in Finnland-Lappland. - Waldhygiene 8: 229–238.
Hölldobler, K., 1960: Systematische Klarstellung zur Ameisenfauna des nordostkarelischen Urwaldes. - Z.angew.Ent. 48: 186–187.
Hung, A.C.F. & Vinson, S.B., 1977: Interspecific Hybridization and Caste Specification of Protein in Fire Ant. - Science 196: 1458–1460.
Jacobson, H., 1939: Die Ameisen des Ostbaltische Gebietes. - Z.Morph.Ökol.Tiere 35: 389–454.
Jensen, T.F., 1977: Some aspects of the ecological energetics of *Formica pratensis* Retz. (Hymenoptera: Formicidae). 80 pp. Thesis, University of Århus, Denmark.
Kilpiäinen, A., Valkeila E., Vesajoki H. & Wuorenrinne, H., 1977: Samettimuurahaisen. - Suomessa Luonnontutkja 81: 129–133.

Kramer, K.U., 1950: Een verwaardloosde Nederlandse *Myrmica* vorm? - Ent.Ber. 13: 97-98.
Kutter, H., 1963: Miscellanea myrmecologica I. Mitt.schweiz.ent.Ges. 36: 129-137.
- 1964: *Formica nigricans* Em. bona species? - Ibid. 37: 132-150.
- 1973: Über die morphologischen Beziehungen der Gattung *Myrmica* zu ihren Satellitengenera *Sifolinia* Em., *Symbiomyrma* Arnoldi und *Sommimyrma* Menozzi (Hymenoptera, Formicidae). - Ibid. 46: 253-268.
- 1977: Hymenoptera Formicidae. - Insecta Helvetica 6, 298 pp. Zürich.
Larsson, Sv.G., 1943: Myrer. - Danmarks Fauna 49, 190 pp. København.
Latreille, P.A., 1798: Essai sur l'histoire des fourmis de la France. 50 pp. Brive.
- 1802a: Histoire naturelle des fourmis. XVI + 445 pp. Paris.
- 1802b: Description d'une nouvelle espèce de fourmi. - Bull.Soc.philomath.Paris 3 (1801-05): 65-66.
- 1804: Nouveau Dictionnaire d'Historie Naturelle etc. Vol. 24. Paris.
- 1805: Histoire naturelle des Crustacés et des Insectes. Vol. 13. Paris.
Linné, C., 1758: Systema naturae. Ed. 10, vol. 1, 824 pp. Holmiae.
- 1761: Fauna svecica. Ed. 2, 578 pp. Stockholmiae.
Lohmander, H., 1949: Eine neue schwedische Ameise. - Opusc.ent. 14: 163-167.
Lomnicki, J., 1925: *Plagiolepis vindobonensis* n.sp. (Hym. Formicidae). - Polskie Pismo ent. 4: 77-79.
Lund, P.V., 1831: Lettre sur les habitudes de quelques fourmis de Brésil, adressée à M. Audouin. - Annls Sci.nat. 23: 113-138.
Mayr, G.L., 1853: Beiträge zur Kenntniss der Ameisen. - Verh.zool.-bot.Ges.Wien 3: 101-114.
- 1855: Formicina austriaca. - Ibid. 5: 273-478.
- 1861: Die europäischen Formiciden. 80 pp. Wien.
- 1862: Myrmecologische Studien. - Verh.zool.-bot.Ges.Wien 12: 649-766.
- 1868: Formicidae Novae americanae coll. a Prof. P.de Strobel. - Annali Soc.nat.Modena 3: 161-181.
Meinert, F., 1861: Bidrag til de danske Myrers Naturhistorie. - K.danske Vidensk. Selsk. Skr. 5: 275-340.
Merisuo, A.K. & Käpylä, M., 1975: Records of a rare ant species, *Camponotus vagus* Rog. (Hym., Formicidae). - Annls Ent.fenn. 41: 140.
Motschulsky, V.v., 1863: Essai d'un catalogue des insectes de l'ile Ceylan. - Bull.Mosc.Soc.Nat. 36: 1-153, 421-532.
Nasonov, N.V., 1889: Contribution to the natural history of the ants of Russia. - Trav.Lab. Zool.Univ.Moscow 4: 1-42.
Nielsen, M.G. & Jensen, T.F., 1975: Økologiske studier over *Lasius alienus* (Först.) (Hymenoptera, Formicidae). - Ent.Meddr 43: 5-16.
Nielsen, M.G., Skyberg, N. & Winther, L., 1976: Studies on *Lasius flavus* (F.) (Hymenoptera, Formicidae): 1 - Population density, biomass, and distribution of nests. - Ent. Meddr 44: 65-75.
Nylander, W., 1846a: Adnotationes in monographiam Formicarum borealium Europae. - Acta Soc.Sci.fenn. 2: 875-944.
- 1846b: Additamentum adnotationum in monographiam Formicarum borealium Europae. - Ibid. 2: 1041-1062.
- 1849: Additamentum alterum adnotationum in monographiam Formicarum borealium. - Ibid. 3 (1848): 25-48.
- 1856: Synopsis des Formicides de France et d'Algérie. - Annls Sci.nat. 5: 51-109.
Oinonen, E.A., 1956: Kallioiden muurahaisista etc. - Acta Ent.fenn. 12: 174-212.

Olivier, G.A., 1791: Fourmis - *Formica*. - Encycl.méthod.Hist.nat. 6: 469–506.
Otto, D., 1958: Zur Schutzwirkung der Waldameisenkolonien gegen Eichenschädlinge. - Waldhygiene 2: 137–142.
- 1962: Die Roten Waldameisen. 151 pp. Wittenberg Lutherstadt.
Pamilo, P., Vepsäläinen, K. & Rosengren, R., 1975: Low allozymic variability in *Formica* ants. - Hereditas 80: 293–296.
Paraschivescu, D., 1972: Pozitia sistematică a speciilor *Formica pratensis* Retz. si *F.nigricans* Em. (Hym., Formicidae). - St. Si Cerc.Biol.Seria Zoologie 24 (6): 527–535.
Passera, L., 1975: Les fourmis hôtes provisoires ou intermediaires des Helminthes. - Annls biol. 14: 227–259.
Petal, J., 1967: Production and the consumption of food in the *Myrmica laevinodis* Nyl. population. - *In* K. Petrusewicz (Ed.): Secondary productivity of terrestrial ecosystems, pp. 841–851. Warsaw.
Petal, J. & Bremeyer, A., 1969: Reduction of wandering spiders by ants in a stellariodeschampsia meadow. - Bull.Acad.pol.Sci.Cl. II Sér.Sci.biol. 17: 239–244.
Pisarski, B., 1961: Studien über die polnischen Arten der Gattung *Camponotus* Mayr (Hymenoptera, Formicidae). - Annls zool. 19: 147–207.
- 1975: Mrowki, Formicoidea. - Kat.Fauny polski 26 (1): 1–85.
Plateaux, L., 1970: Sur le polymorphisme social de la fourmi *Leptothorax nylanderi* Förster. - Annls Sci.nat.Zool. 12. serie, 12: 373–478.
Poldi, B., 1962: Alcuni appuncti su una rara formica (*Lasius bicornis* Först.) nuova per la Sardegna. - Studi sassar. 9: 509–516.
Pontin, A.J., 1978: The numbers and distribution of subterranean aphids and their exploitation by the ant *Lasius flavus* (Fabr.). - Ecol.Ent. 3: 203–207.
Retzius, A.J., 1783: Caroli Lib.Bar. de Geer et species insectorum. - 220 pp., Leipzig.
Roger, J., 1859: Beiträge zur Kenntniss der Ameisenfauna der Mittelmeerländer. - Berl.ent.Z. 3: 225–259.
Rosengren, R., 1969: Notes regarding the growth of a polycalic nest system in *Formica uralensis* Ruzsky. - Notul.ent. 49: 211–230.
- 1971: Route fidelity, visual memory and recruitment behavior in foraging wood ants of the genus *Formica* (Hymenoptera, Formicidae). - Acta zool.fenn. 133: 1–106.
- 1977a: Foraging strategy of wood ants (*Formica rufa* group). I. Age polyethism and topographic traditions. - Acta zool.fenn. 149: 1–30.
- 1977b: Foraging strategy of wood ants (*Formica rufa* group). II. Nocturnal orientation and diel periodicity. - Ibid. 150: 1–30.
Ruzsky, M.D., 1895: Faunistische Untersuchungen in östlichen Russland. - Kazan Soc.Nat. Trans. 28 (No. 5): 64 pp. (In Russian).
- 1896: Verzeichniss der Ameisen des östlichen Russland und Uralgebirges. - Berl.ent.Z. 44: 67–74.
- 1905: Formicarii Imperii Rossici. - Arb. naturf.Ges.Kais.Univ.Kasan 38.
Sadil, J., 1951: A revision of the Czechoslovak forms of the genus *Myrmica* Latr. (Hym.). - Sb.ent.Odd.nár.Mus.Praze 27: 233–278.
Santschi, F., 1938: Notes sur quelques *Ponera*. - Bull.Soc. ent.Fr. 43: 78–80.
Schenck, C.F., 1852: Beschreibung nassauischer Ameisenarten. - Jahrb.Ver.Naturkd.Nassau 8: 3–149.
Scopoli, J.A., 1763: Entomologia Carniolica. - 451 pp. Vindobonae.
Skøtt, C., 1971: Nye danske fund af myren *Ponera punctatissima* Roger (Hym., Formicidae). - Ent.Meddr 39: 44–47.

- 1973: Nye danske fund af myrerne *Stenamma westwoodii* Westw. og *Myrmecina graminicola* Latr. (Hym., Formicidae). - Flora og fauna 79: 11.
Smith, F., 1851: List of the specimens of British animals in the collection of the British Museum: 6 - Hymenoptera Aculeata. - 134 pp., London.
Stärcke, A., 1942: Drie nog onbeschreven Europeesche miervormen. - Tijdschr.Ent. 85: XXIV-XXIX.
Stitz, H., 1939: Hautflügler oder Hymenoptera. I: Ameisen oder Formicidae. - Tierwelt Dtl. 37, 428 pp. Jena.
Sudd, J.H., 1966: An Introduction to the Behaviour of Ants. 200 pp. Edward Arnold, London.
Taylor, R.W., 1967: A monographic revision of the ant genus *Ponera* (Latreille) (Hymenoptera: Formicidae). - Pacif. Insects Mon. 13, 112 pp.
Waldén, H.W., 1964: Ett jättelikt bo av blanksvart trädmyra *Lasius fuliginosus*. - Göteborgs Mus.årstryck 1964: 20-27.
Westwood, J.O., 1840: An introduction to the modern classification of insects. Vol. 1, 158 pp. Longman, London.
- 1841: Observations on the genus *Typhlopone*, with description of several exotic species of ants. - Ann.Mag.nat.Hist. 6: 81-89.
Wilson, E.O., 1955: A monographic revision of the ant genus *Lasius*. - Bull.Mus.comp.Zool. Harv. 113: 1-199.
- 1971: Insect Societies. 548 pp. Belknap Prerss, Harvard University, Cambridge.
Wilson, E.O., Carpenter, F.F. & Brown, W.L., 1967: The first mesozoic ants. - Science 157: 1038-1040.
Wing, M.W., 1968: Taxonomic revision of the nearctic genus *Acanthomyops* (Hymenoptera: Formicidae). - Mem.Cornell Univ.agric.Exp.Stn 405: 1-173.
Wuorenrinne, H., 1974: Suomen kekomuurahaisten (*Formica rufa* coll.) ekologiasta ja levinneisyydestä. - Silva fenn. 8: 205-214.
- 1978: The influence of collection of ant pupae upon ant populations in Finland (Hymenoptera, Formicidae). - Notul.ent. 58: 5-11.
Wuorenrinne, H. & Vepsäläinen, K., 1976: Effects of environmental splitting by urbanisation on the species of the *Formica rufa* group. - Proc. II Internat. Symposium, IUSSI, Warsaw. pp. 69-78.
Yarrow, I.H.H., 1954: The British ants allied to *Formica fusca* L. (Hym., Formicidae). - Trans. Soc.Br.Ent. 11: 229-244.
- 1955a: The British ants allied to *Formica rufa* L. (Hym., Formicidae). - Ibid. 12: 1-48.
- 1955b: The type species of the ant genus *Myrmica* Latreille. - Proc.R.ent.Soc.Lond. (B) 24: 113-115.
- 1968: *Sifolinia laurae* Emery, 1907. A workerless parasitic ant new to Britain. - Entomologist 101: 236-40.
Zetterstedt, J.W., 1840: Insecta Lapponica. vi + 1,140 pp. Lipsiae.

Index

Synonyms are given in italics. The number in bold refers to the main treatment of the taxon.

Author's address:
C. A. Collingwood
Agricultural Development and Advisory Service
Government Buildings, Lawnswood
Leeds LS16 5PY, England

SWEDEN

Sk.	Skåne	Vrm.	Värmland
Bl.	Blekinge	Dlr.	Dalarna
Hall.	Halland	Gstr.	Gästrikland
Sm.	Småland	Hls.	Hälsingland
Öl.	Öland	Med.	Medelpad
Gtl.	Gotland	Hrj.	Härjedalen
G. Sand.	Gotska Sandön	Jmt.	Jämtland
Ög.	Östergötland	Ång.	Ångermanland
Vg.	Västergötland	Vb.	Västerbotten
Boh.	Bohuslän	Nb.	Norrbotten
Dlsl.	Dalsland	Ås. Lpm.	Åsele Lappmark
Nrk.	Närke	Ly. Lpm.	Lycksele Lappmark
Sdm.	Södermanland	P. Lpm.	Pite Lappmark
Upl.	Uppland	Lu. Lpm.	Lule Lappmark
Vstm.	Västmanland	T. Lpm.	Torne Lappmark

NORWAY

Ø	Østfold	HO	Hordaland
AK	Akershus	SF	Sogn og Fjordane
HE	Hedmark	MR	Møre og Romsdal
O	Opland	ST	Sør-Trøndelag
B	Buskerud	NT	Nord-Trøndelag
VE	Vestfold	Ns	southern Nordland
TE	Telemark	Nn	northern Nordland
AA	Aust-Agder	TR	Troms
VA	Vest-Agder	F	Finnmark
R	Rogaland		

n northern s southern ø eastern v western y outer i inner

FINLAND

Al	Alandia	Kb	Karelia borealis
Ab	Regio aboensis	Om	Ostrobottnia media
N	Nylandia	Ok	Ostrobottnia kajanensis
Ka	Karelia australis	ObS	Ostrobottnia borealis, S part
St	Satakunta	ObN	Ostrobottnia borealis, N part
Ta	Tavastia australis	Ks	Kuusamo
Sa	Savonia australis	LkW	Lapponia kemensis, W part
Oa	Ostrobottnia australis	LkE	Lapponia kemensis, E part
Tb	Tavastia borealis	Li	Lapponia inarensis
Sb	Savonia borealis	Le	Lapponia enontekiensis

USSR

Vib	Regio Viburgensis	Kr	Karelia rossica	Lr	Lapponia rossica

Printed in the United States
By Bookmasters